Lab Manı

to Accompany

Introduction to Telecommunications Networks

Paul Barufaldi

Springfield Technical Community College

THOMSON

DELMAR LEARNING™

Australia Canada Mexico Singapore Spain United Kingdom United States

Lab Manual to accompany Introduction to Telecommunications Networks
Paul Barufaldi

Vice President, Technology and Trades SBU:
Alar Elken

Editorial Director:
Sandy Clark

Senior Acquisitions Editor:
Gregory L. Clayton

Development:
Dawn Daugherty

Marketing Director:
Maura Theriault

Channel Manager:
Fair Huntoon

Marketing Coordinator:
Brian McGrath

Production Director:
Mary Ellen Black

Production Manager:
Larry Main

Production Coordinator:
Sharon Popson

Senior Project Editor:
Christopher Chien

Art/Design Coordinator:
Francis Hogan

Technology Project Manager:
David Porush

Printed in the United States of America
1 2 3 4 5 XX 06 05 04 03 02

For more information contact
Delmar Learning
Executive Woods
5 Maxwell Drive, PO Box 8007
Clifton Park, NY 12065-8007

Or find us on the World Wide Web at
http://www.delmarlearning.com

Library of Congress Catalog Card Number: 2002075054

NOTICE TO THE READER

Contents

Preface

The experiments in this manual are designed to accompany *Introduction to Telecommunications Networks* by Gordon F. Snyder and reinforce many of the concepts in the text. Each experiment provides hands-on experience in basic telecommunications tasks such as terminating both copper and fiber cable ends.

As a prerequisite, students should have a basic knowledge of AC and DC circuit theory and be able to use test equipment to measure voltage, current, resistance, frequency, and capacitance. Students should also be able to solder, strip wire, and use hand tools to disassemble and reassemble computer equipment. Also needed is a basic knowledge of computers and the Windows™ operating system.

Most of the equipment and materials used in these experiments are readily available through electrical supply houses, computer stores, and catalog companies.

The following is a list of vendors through which equipment can be purchased:

Handtools and telecommunications test equipment

http://www.jensentools.com/

TCM-100 trainer

http://www.elexp.com/tst_m100.htm

Fiber termination equipment

http://www.fiberinstrumentsales.com

The experiments follow the book contents in order with the exception of Experiment 1—Direct Cable Connection. I like to do this experiment the first week to provide the students with experience using handtools, soldering, using the multimeter for continuity checks, and interfacing with the computer. It also allows the students to work in teams.

Acknowledgments

Delmar Learning and the author gratefully acknowledge the contributions of the review panel, whose valuable comments helped shape this laboratory manual. The review panel included:

Mike Beaver
University of Rio Grande
Rio Grande, OH

Joe Gryniuk
Lake Washington Technical College
Kirkland, WA

Ryan McCaigue
DeVry University
Phoenix, AZ

Judson Miers
DeVry University
Kansas City, MO

NCTT

The National Center for Telecommunications Technologies (NCTT: http://www.nctt.org) is a National Science Foundation (NSF: http://www.nsf.gov) sponsored Advanced Technological Education (ATE) Center established in 1997 by Springfield Technical Community College (STCC: http://www.stcc.edu) and the NSF. All material produced as part of the NCTT textbook series is based on work supported by STCC and the NSF under Grant Number DUE 9751990.

NCTT was established in response to the telecommunications industry and the worldwide demand for instantly accessible information. Voice, data, and video communications across a worldwide network are creating opportunities that did not exist a decade ago, and preparing a workforce to compete in this global marketplace is a major challenge for the telecommunications industry. As we enter the twenty-first century, with even more rapid breakthroughs in technology anticipated, education is the key and NCTT is working to provide the educational tools employers, faculty, and students need to keep the United States competitive in this evolving industry.

We encourage you to visit the NCTT Web site at http://www.nctt.org along with the NSF Web site at http://www.nsf.gov to learn more about this and other exciting projects. Together we can explore ways to better prepare quality technological instruction and ensure the globally competitive advantage of America's telecommunications industries.

Materials and Equipment

5-Minute Epoxy
62.5/125 Jacketed Fiber-Optic Cable
Alcohol Pad
Analog Modem
ASCII Table
Assorted Serial Cables
Audio-Frequency Generator
AWG Table
BNC Connectors
Butt Set
Capacitance Meter
Capacitance Substitution Box
Capacitor, 0.01 μF
Capacitor 0.1 μF
Category-5 Patch Cable
Category-3 Wire
Category-5 Wire
Cisco Catalyst 1900-Series Switch
Cleave Tool
Coaxial Cable, RG-58/U
Coaxial Cable, RG-59
Coaxial Wire Stripper
Computer with a Free Serial Port
Computer with Internet Access
Computer with Network Card
Computers with Windows™ Operating System
Console (Rollover) Cable
Crossover Cable
DB-9 Female Connector
DC Power Supply
Decade Resistance Box
Double-Sided Tape
Ethereal Protocol Analyzer
Ethernet Switch
F Connectors for RG-59
Fiber Cable Stripper
Fiber Crimp Tool
Fiber Stripper
Flat Telephone Cable, 4-Conductor
Fluke 620 LAN Cable Meter
Fox and Hound
F-Type Crimping Tool
Hub
HyperTerminal Program
Kevlar Cutting Scissors
Microscope
Multi-Grit Fiber Polishing Paper, 5 μm, 1 μm, 0.3 μm
Multimeter
NIC
Oscilloscope, Dual-Trace
Oscilloscope Storage
Polishing Bushing
Polishing Pad
Polishing Plate
Punch-Down Tool
Resistor, 1000 Ω
RJ-11 Crimper
RJ-11 Jack
RJ-11 Plug
RJ-45 Crimping Tool
RJ-45 Plug
RJ-45 to DB-9 Female Terminal Adapter
SC Fiber-Optic Connector Kit
Short Haul Modem
Solder
Soldering Iron
TCOM 100 Trainer
Telephone
Telnet Program
Web Browser
Wire Cutters
Wire Strippers

Experiment 1

Direct Cable Connection

Name ______________________________ Class ____________ Date ______________

Objectives Upon completion of this experiment, you should be able to:

- Connect two computers together through the serial port.
- Construct a null modem cable.
- Transfer files using the Direct Cable Connection program.

Materials and Equipment

DB-9 Female Connector (2)
Computer with Windows™ 98 and a Free Serial Port (2)
8-Conductor Wire
Soldering Iron
Solder
Multimeter
Wire Cutters
Wire Strippers

Introduction

One way for you to transfer files from one computer to another is to use the Windows Direct Cable Connection program and a null modem cable using the standard serial port on your computer. In this lab you will build a null modem cable, test the individual pins for continuity, and use the cable to transfer files between two computers.

Procedures

1. Building a null modem cable.

   A. Use a piece of 8-conductor wire that is long enough to connect the two computers you will be using.

   B. Solder both DB-9 female connectors using the pinout diagram in Table 1-1.

   C. Pin numbers are on both sides of the connector and can be hard to see, so draw a pinout diagram of the side you will be soldering.

End 1 Pin Number		End 2 Pin Number	Wire Color
1 and 6	*to*	4	
2	*to*	3	
3	*to*	2	
4	*to*	1 and 6	
5	*to*	5	
7	*to*	8	
8	*to*	7	

Table 1-1

D. Fill in the color of your wire in Table 1-1 to aid in assembly.

E. Check each pin for proper continuity and make sure there are no shorts between any pins that are not supposed to be connected.

Verified: ___________

2. Communicating between computers. *The procedures in this step should be run on both of your computers.*

A. Check to see if the Direct Cable Connection program is installed on your computers.

(1) Click **Start/Programs/Accessories/Communications.**

(2) If Direct Cable Connection is listed, then it is installed; if it is not listed, then go to step 2B.

B. If the Direct Cable Connection program is not listed you need to install it.

(1) Click **Start/Settings/Control Panel.** Double click on **Add/Remove Programs.** Click the **Windows Setup** tab. Double click the **Communications** icon. Check the **Direct Cable Connection** box and click **OK.**

(2) Reboot the computer.

C. Set up File and Sharing.

(1) Click **Start/Settings/Control Panel/Network.** Open the network by double clicking on the **Network** icon.

(2) To enable file sharing, select **I want to be able to allow others access to my files** under File and Print Sharing and then click **OK.** See Figure 1-1.

(3) Reboot the computer.

D. Set up the sharing of files.

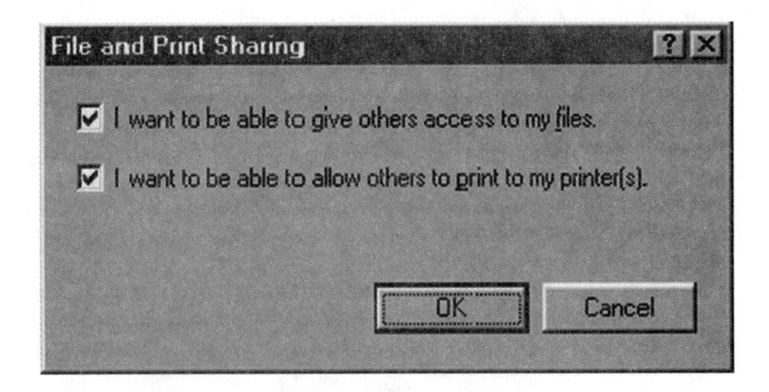

Figure 1-1 File and Print Sharing

(1) Launch Windows Explorer by either clicking on **Start/Programs/Windows Explorer** or alternate clicking on **Start/Explore.**

(2) Highlight the **C drive** and then select **Sharing.**

(3) Under the **Sharing** tab select the **Shared As** radio button and accept the default share name. See Figure 1-2.

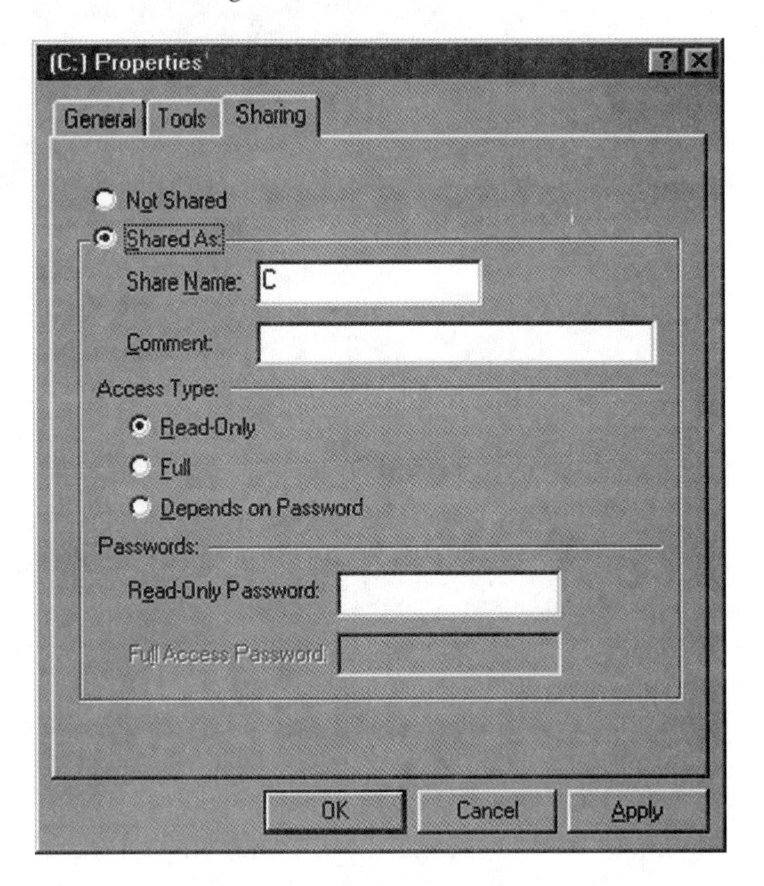

Figure 1-2 C: Properties

E. Select one computer to be the host and the other to be the guest.

F. Set up the host computer.

(1) Click on **Start/Programs/Accessories/Communications**. Click on **Direct Cable Connection.**

(2) In the Direct Cable Connection box, choose the option for **Host**. See Figure 1-3.

(3) Choose the communications port.

(4) Click **Next** and your host computer will wait for the guest computer to connect to it.

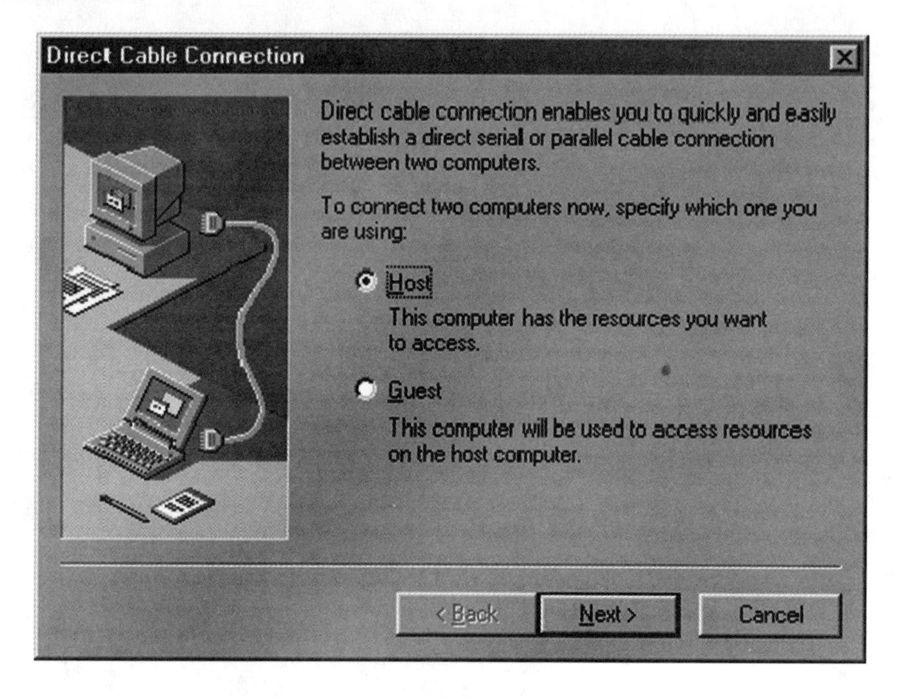

Figure 1-3 Direct Cable Connection—Host

G. Set up the guest computer.

(1) Click on **Start/Programs/Accessories/Communications**. Click on **Direct Cable Connection**.

(2) In the Direct Cable Connection box, choose the option for **Guest**. See Figure 1-4.

(3) Choose the communications port.

(4) Click **Next** and the guest computer will look for the host computer.

3. Once the guest computer and the host computer are connected you will be able to browse the shared drive through Network Neighborhood and be able to copy files.

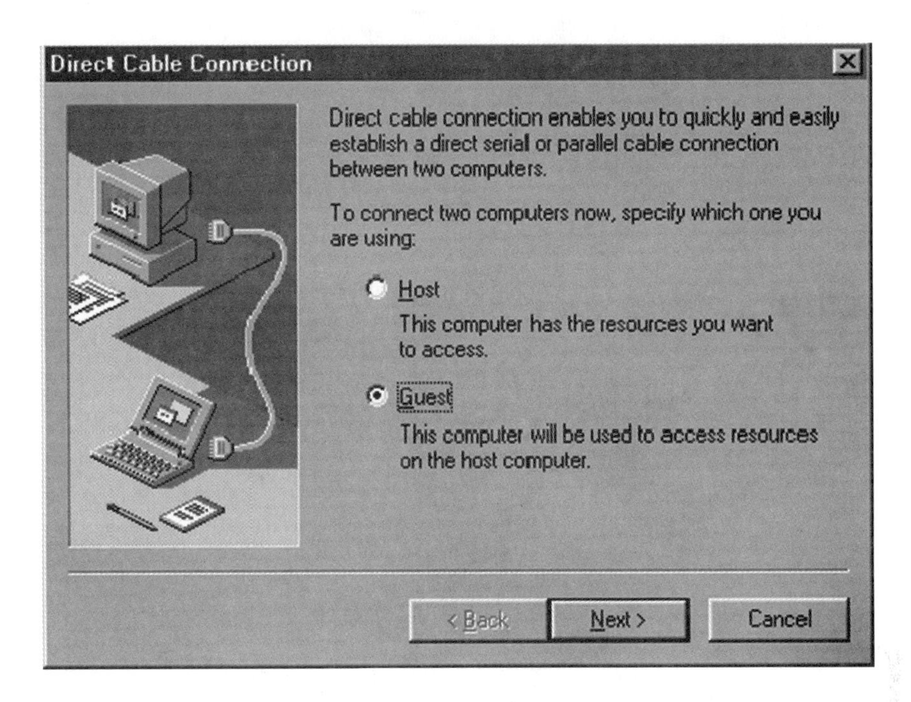

Figure 1-4 Direct Cable Connection—Guest

Experiment

2

Punch-Down Tool and Punch-Down Blocks

Name ______________________ Class __________ Date ____________

Objectives Upon completion of this experiment, you should be able to:

- Identify the two types of punch-down blocks.
- Identify the two types of punch-down blades.
- Install and test wire for continuity.
- Identify *tip* and *ring*.

Text Reference Snyder, *Introduction to Telecommunications Networks*
Chapter 2

Materials and Equipment
Punch-Down Tool
Wire Strippers
Wire Cutters
Category-3 Wire
TCOM 100 Trainer
Telephone
Multimeter
RJ-11 Jack

Introduction

Installing wires, repairing broken wires, making custom wires, and testing the continuity of wires are common tasks for telecommunications technicians. Insulation Displacement Connection (IDC) termination is the recommended method of copper termination recognized by ANSI/TIA/EIA-568-A for UTP cable terminations. Terminations are made on punch-down blocks (type 66 and type 110) with a punch-down tool specific for each type of block. Punch-down tools may have interchangeable bits (one for each type of block) and can be either cutting or non-cutting.

Procedures

1. All connections are made with power to the trainer *turned off!*

2. Cut two one-foot pieces of Category-3 telecommunications wire. I will use a red and green wire as an example.

3. Examine the punch-down tool; notice that each blade has both a cutting and a non-cutting side. We will utilize the non-cutting side. See Figure 2-1.

   Verified: __________

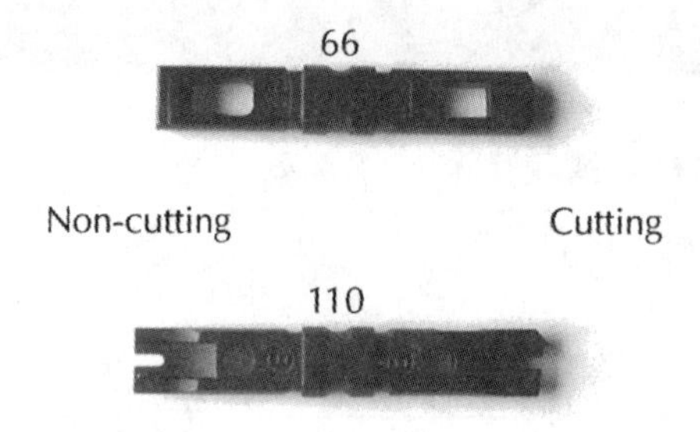

Figure 2-1 66 and 110 Punch-Down Blades

4. Using the 110 blade, punch down one end of cable one onto the PB6 blue terminal with the green wire to the left side and the red wire to the right side. See Figure 2-2. *Do not* strip the insulation off the individual wires.

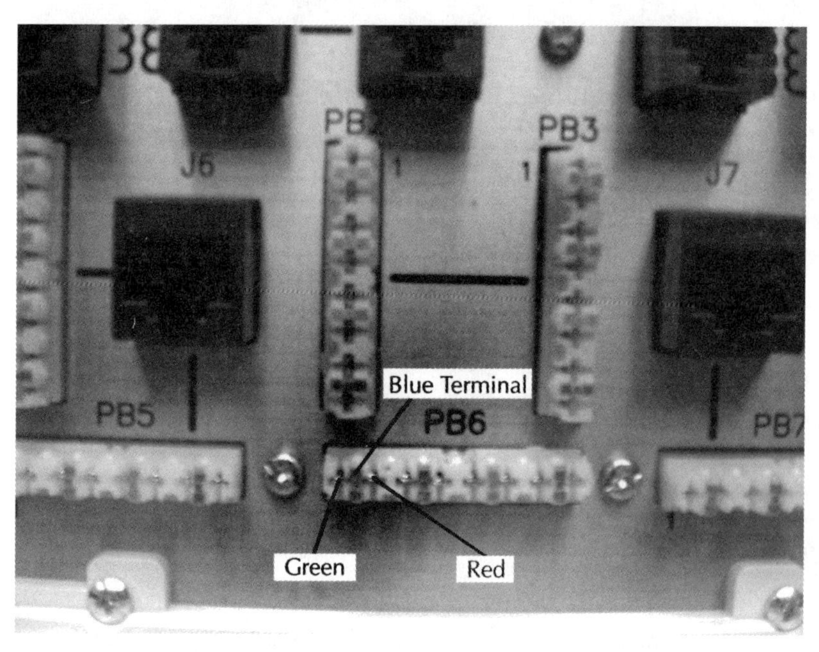

Figure 2-2 PB6 Blue Terminal

5. Using the 66 blade, punch down the other end of the cable to the two outside pins of the 66 block with red on the top and green on the bottom. See Figure 2-3.
6. Remove the wires just installed and repeat steps 4 and 5 using the cutting side of the punch-down tool.
7. Punch down another wire from the inside pins of the 66 block. Match the colors to the breadboard and utilize the two vertical connections.
8. Connect an RJ-11 jack to the breadboard using the same vertical connection row used in step 5. Connect the ring (red) to pin 2 and tip (green) to pin 3. See Figure 2-4.
9. Check the continuity from the RJ-11 jack to the connections on PB6.

   Verified: ___________

Figure 2-3 66 Block

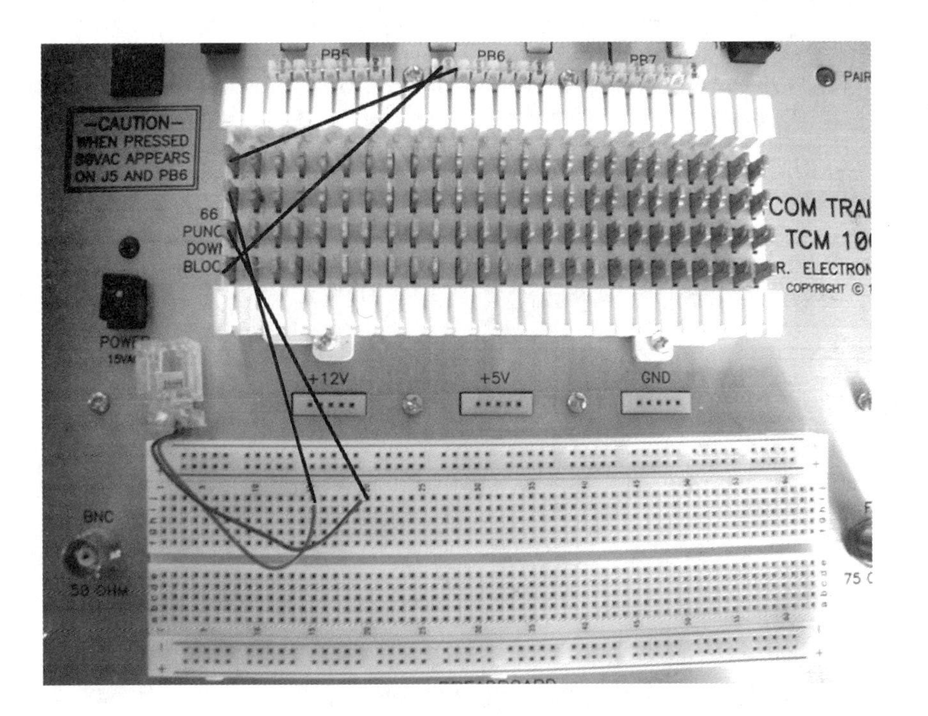

Figure 2-4 RJ-11 on Breadboard

10. Does pin 2 on the RJ-11 jack connect to the left side of the PB6 blue terminal?

    Verified: ___________

11. Does pin 3 on the RJ-11 jack connect to the right side of the PB6 blue terminal?

    Verified: ___________

12. Draw a schematic of the circuit you just constructed.

13. Plug a phone into the RJ-11 jack connected to the breadboard.

14. Power up the trainer.

15. Place the phone in the off-hook position and press the dial-tone button on the trainer.

16. Do you hear a dial tone?

    Verified: ___________

17. If you do not hear a dial tone, then check the wiring.

Questions

1. What is the difference between the on-hook and off-hook positions on a phone?

____

____

____

2. Where would you expect to find a 66-type punch-down block?

____

____

____

3. Where would you expect to find a 110-type punch-down block?

____

____

____

4. Which did you find easier to punch down, the 66- or 110-type connection?

____

____

____

5. Why is it unnecessary to strip the wire ends before insertion onto the blocks?

____

____

____

Experiment 3

Telephone Set Function

Name ______________________________ Class __________ Date ____________

Objectives Upon completion of this experiment, you should be able to:

- Make a phone cable.
- Understand the difference between straight-pinned and cross-pinned cables.
- Measure both on-hook and off-hook signals.
- Measure the ring signal.
- Understand *tip* and *ring.*
- Measure loop currents.
- View the DTMF signals on an oscilloscope.

Text Reference Snyder, *Introduction to Telecommunications Networks*
Chapter 2

Materials and Equipment
Category-3 Wire, 4 Pair
Flat Telephone Cable, 4-Conductor
RJ-11 Jack
RJ-11 Plugs (2)
RJ-11 Crimper
Multimeter
TCOM 100 Trainer
Dual-Trace Oscilloscope
Punch-Down Tool

Introduction

Making patch cables is a common task for a telecommunications technician. Patch cables can be straight-pinned, where each pin of the connector matches the corresponding pin at the other end of the cable, or cross-pinned, where the pin at one end is reversed from that at the other end. To understand how a phone set operates, you can measure the voltage and current values with a multimeter as well as view the audio and tone signals with an oscilloscope.

Procedures

1. Research to determine the following:

   A. On-hook voltage __________.

   B. Off-hook voltage __________.

C. On-hook current ___________.

D. Off-hook current ___________.

E. Ring voltage ___________.

2. Make a 3-foot cross-pinned patch cable using RJ-11 plugs and flat cable. See Table 3-1 and Figure 3-1 for the pin connections. Cross-pinned cables are used to connect a phone to a modular jack on the wall.

End 1	Color	End 2
Pin 1	Black	Pin 4
Pin 2	Red	Pin 3
Pin 3	Green	Pin 2
Pin 4	Yellow	Pin 1

Table 3-1

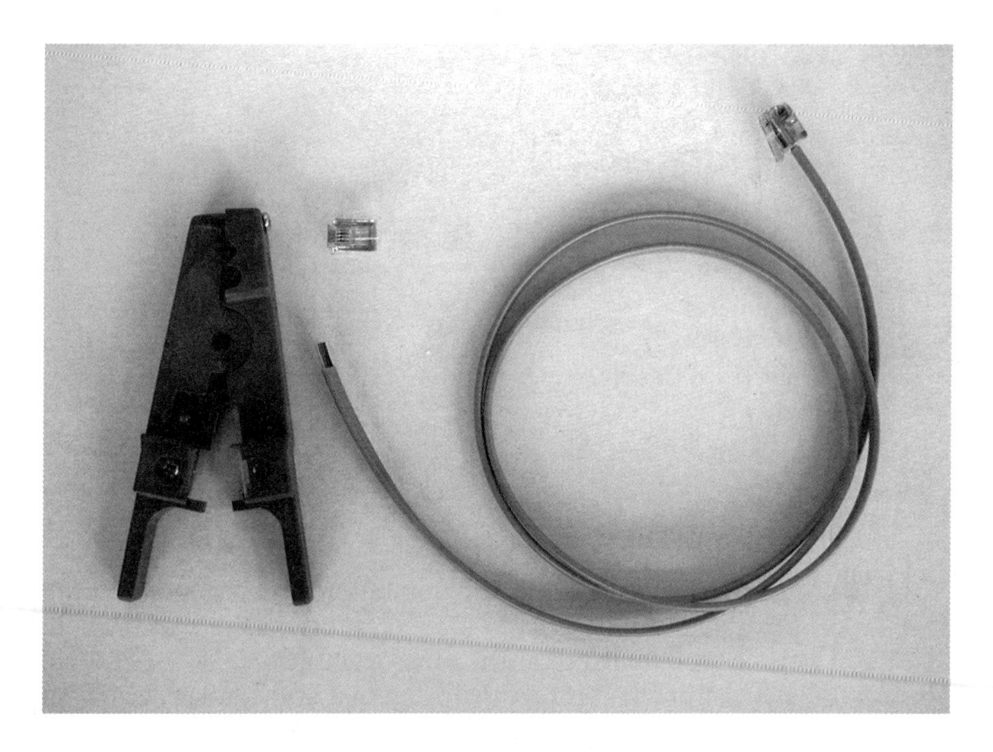

Figure 3-1 Cross-Pinned Cable with Wire Stripper

3. Make sure that the wire ends can be seen at the end of the plug before crimping.

   Verified: ___________

4. Identify pin 1 by holding the connector with the pins up. Clip down and view it from the back of the plug where the wire goes in. Pin 1 is on the far left.

5. Draw a diagram of the patch cable identifying the pin numbers with the associated color of the wire connected to it.

6. Punch down a wire pair using the Category-3 wire from the 110-block PB6 brown terminal. The green wire should connect to T (left of brown) and the red to R (right of brown) on the left outside pins of the 66 block. See Figure 3-2.

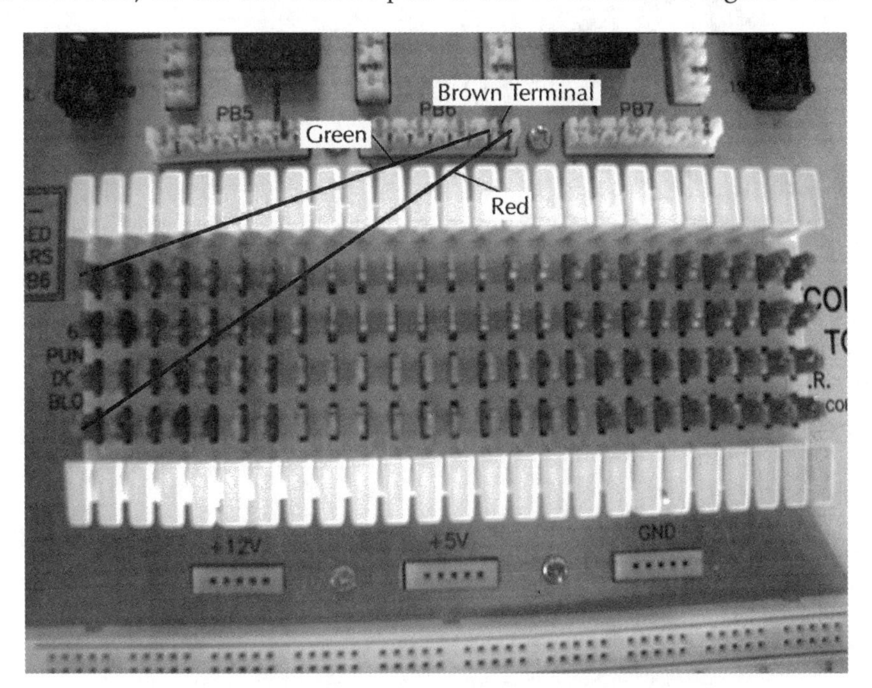

Figure 3-2 PB6 Brown Terminal

7. With your patch cable, plug the phone into jack J5 on the trainer.
8. Power up the trainer.
9. Measure on-hook voltage at the 66 block ___________.
10. Is the reading the same as 1A?

    Verified: ___________
11. Measure off-hook voltage at the 66 block ___________.
12. Is the reading the same as 1B?

    Verified: ___________
13. Disconnect power from the trainer.

14. Punch down another wire from the inside pins of the 66 block. Match the colors to the breadboard and utilize two vertical connections. See Figure 3-3.

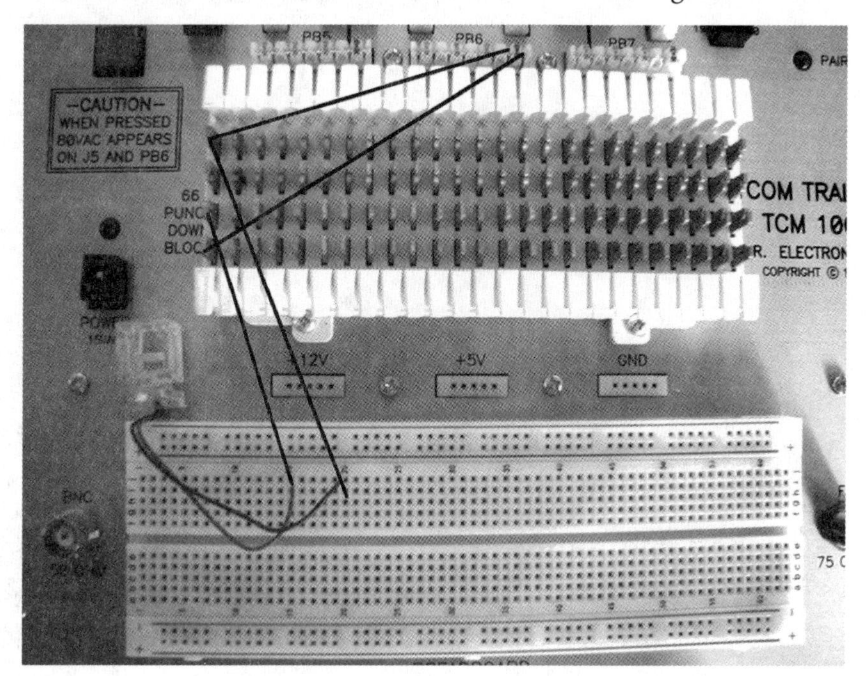

Figure 3-3 66 Block to Breadboard

15. Attach an RJ-11 jack to the breadboard using the same vertical connection row used in step 14 and connect the ring (red) to pin 2 and tip (green) to pin 3, as depicted in Figure 3-3.

16. Plug the phone into the jack on the breadboard with the patch cable you made in step 2.

17. Check the continuity from the RJ-11 jack to the connections on PB6. Draw a schematic of the circuit you just constructed. Do not forget to draw in the phone's hook switch.

18. Power up the trainer.

19. Measure on-hook current ___________.

20. Is the reading the same as 1C?

    Verified: ___________

21. Measure off-hook current ___________.

22. Is the reading the same as 1D?

    Verified: ___________

23. How did you measure the current?

____

____

____

____

____

24. Determine the off-hook resistance using Ohm's Law ___________.

25. Plug the phone into jack J5 on the trainer.

26. Connect the AC voltmeter to the inner pins on the 66 block.

27. Press the ring signal and measure the ring voltage ___________.

28. Did the phone ring?

    Verified: ___________

29. Is the reading the same as 1E?

    Verified: ___________

30. Remove the phone from jack J5.

31. Plug the phone into the jack on the breadboard.

32. Connect an oscilloscope to the inner pins on the 66 block.

33. Press the ring signal and measure ring voltage ___________.

34. Is the voltage the same as in step 27?

    Verified: ___________

35. Talk into the mouthpiece and observe the waveform on the oscilloscope.

36. Use the oscilloscope to view and measure the DTMF frequencies. Frequencies are shown in Figure 3-4.

697 Hz	1	2	3
770 Hz	4	5	6
852 Hz	7	8	9
941 Hz	*	0	#
	1209 Hz	1336 Hz	1477 Hz

Figure 3-4 DTMF Pad

37. Have another person press a button on the phone and see if you can tell what the number is by measuring the frequency.

Questions

1. What is *DTMF*?

2. What is *REN*?

3. What happens when multiple phones are connected to the same phone line?

4. What is the difference between *tip* and *ring*?

5. Explain the actions that take place when a phone goes into the off-hook position?

6. Where do DTMF phones get the power to generate an electronic signal?

Experiment 4

Telephone Wiring Troubleshooting

Name ______________________ Class __________ Date ____________

Objectives Upon completion of this experiment, you should be able to:

- Demonstrate the use of a fox-and-hound set.
- Demonstrate the use of a butt set.
- Troubleshoot wire-pair tracing.
- Document wire-pair installation.

Text Reference Snyder, *Introduction to Telecommunications Networks*
Chapter 2

Materials and Equipment
Butt Set
Fox and Hound
TCOM 100 Trainer
Category-3 Wire, 4 Pair
Punch-Down Tool

Introduction

Troubleshooting telephone problems includes wiring diagrams. Because wiring between the PBX and the telephone set can travel great distances within buildings through multiple wiring closets, it is necessary to be able to trace the wire pair to aid in troubleshooting. The butt set looks somewhat like a telephone handset with the dial pad built into it. It comes with a cord set with clips for easy access to the tip and ring signals. The butt set has a speaker that will allow you to listen for different telephone signals such as a ring signal and dial tone. The fox-and-hound set can be used to quickly identify wire pairs.

Procedures

1. Install the wire.

   A. Using three TCOM trainers, label them 1, 2, and 3.

   B. Place the trainers in three different rooms.

   C. From the 66 block on trainer 1, punch down a wire using the outer connectors with 4 or more pairs. See Figure 4-1.

   D. Document the wire color and location on the block.

   E. Run wire 1 from trainer 1 to trainer 2 and punch down.

Figure 4-1 66 Block—4 Wires

F. Punch down another wire on the 66 block using the inner connectors of trainer 2.

G. Document the wire color and location on the block.

H. Run wire 2 from trainer 2 to trainer 3 and punch down.

I. Document the wire color and location on the block.

2. Test the connection with a fox and hound.

   A. Using the tone generator (fox) (see Figure 4-2) and the inductive pickup (hound) (see Figure 4-3), test each wire pair from trainer 1 to trainer 2, from trainer 2 to trainer 3, and from trainer 1 to trainer 3.

   B. Document the results for each wire pair. If the wire pairs do not trace, punch them down again and retest.

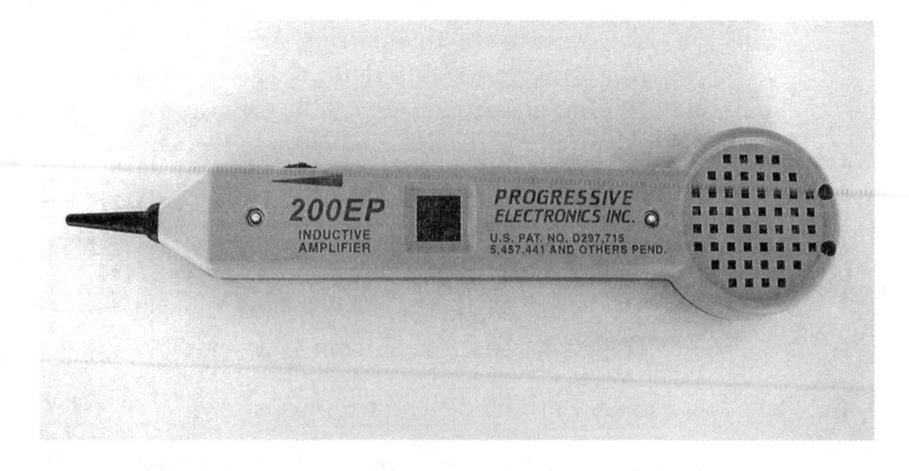

Figure 4-2 Fox

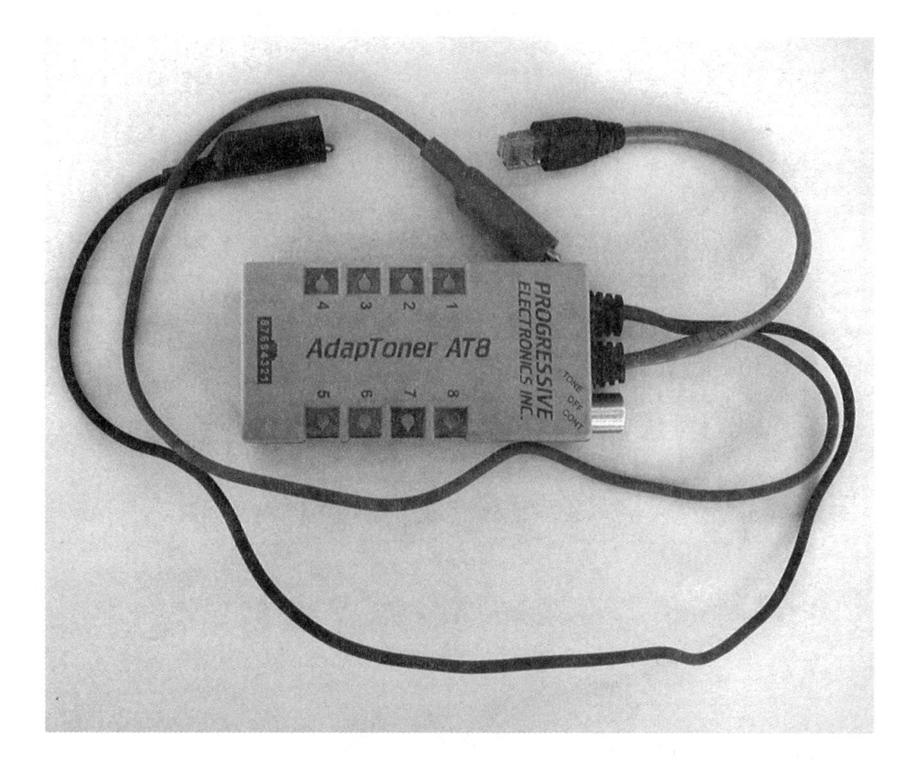

Figure 4-3 Hound

3. Test with a butt set.

   A. Punch down a wire pair using the Category-3 wire from the 110-block PB6 blue terminal. The green wire should connect to T (left of blue) and the red to R (right of blue) on left inside pins of the wire pair in the first row of the 66 block. See Figure 4-4. Do this on trainer 1 and trainer 3.

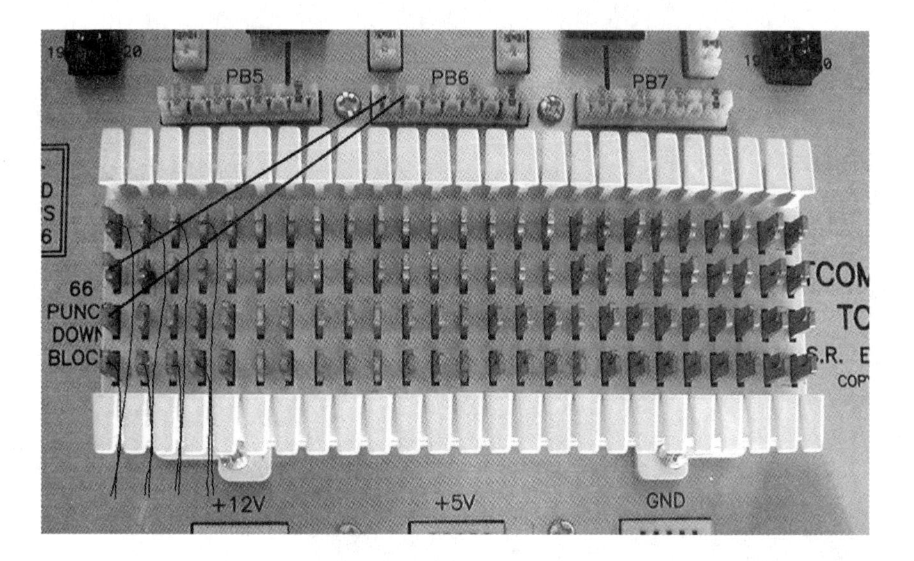

Figure 4-4 PB6 Blue Terminal

   B. Plug the phone into jack J2 on both trainers.

   C. Power up the trainer.

D. On trainer 2, attach the butt set to the inner terminals of the first row of the 66 block. See Figure 4-5.

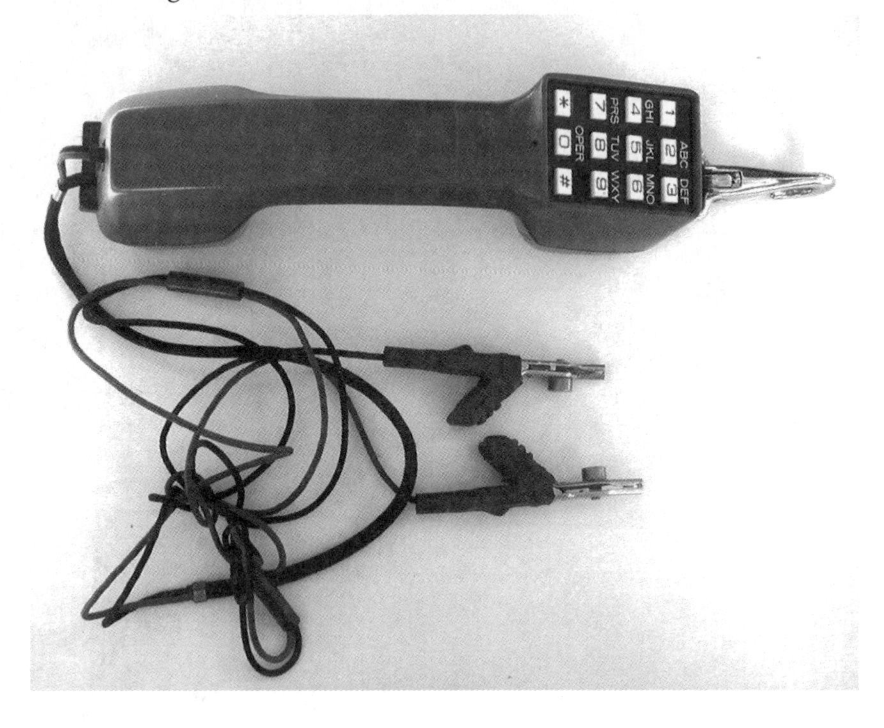

Figure 4-5 Butt Set

E. Have the person on trainer 1 push the dial-tone signal and verify that you can hear it on the butt set.

Verified: ___________

F. Have the person on trainer 1 dial some numbers and verify that you hear the tones.

Verified: ___________

G. Repeat steps E and F for trainer 3.

H. Verify that trainer 1 can talk to trainer 3.

Verified: ___________

I. Have the instructor create a problem in your phone system and then test the system to see if you can find and repair the fault.

Questions

1. List as many faults as you can that could be associated with one phone not communicating with another phone in a PBX.

2. How long did it take you to find and fix the problem your instructor injected into your phone system?

3. Which troubleshooting tools did you use to find the problem?

Experiment 5

Wire Resistance

Name ______________________ Class ____________ Date ____________

Objectives Upon completion of this experiment, you should be able to:

- Calculate the resistance of a coil of wire based on wire length, type, and cross-sectional area.
- Calculate the resistance of a coil of wire based on the American Wire Gauge (AWG).
- Calculate the resistance of a coil of wire based on Ohm's Law.
- Measure the resistance of a coil of wire.

Text Reference Snyder, *Introduction to Telecommunications Networks*
Chapter 2, Section 1

Materials and Equipment
Category-3 Wire, 1000 Feet or Greater
Multimeter
AWG Table
DC Power Supply

Introduction

Wire lengths are critical to the DC resistance of a local loop. As the line length increases, signals become weaker and because line length can cause minor problems in analog transmission, it can also cause major problems in digital transmissions.

Procedures

1. You will need to know the length of the wire loop of which you are measuring the resistance.
2. Determine the wire gauge and area in circular mils of the wire being used.

   Gauge ____________

   Circular Mil Area ____________
3. Short out one end on the wire pair to be measured.

4. Calculate the resistance of a wire pair in a coil of wire using the following formula:

$$R_X = e \text{ (resistvity)} \cdot \text{Length (feet)/Area (circular mils)}$$

Resistivity of copper = 10.4

Circular Mils = C.M. = (diameter)^2/1000

R_x = __________

5. Determine the resistance per 1000 feet of wire using the gauge value from Procedure step 2 and Table 5-1.

Ω/1000 feet – __________

Wire Gauge	Circular Mil Area	Ohms per 1000 Feet
19	1288.0	9.5
22	642.4	19
24	404.0	30.2
26	254.1	48

Table 5-1

6. Calculate the resistance by the gauge value in Table 5-1.

$$R = \Omega/1000 \text{ feet} \cdot \text{length (feet)}$$

R = __________

7. Calculate the resistance using Ohm's Law.

Ohm's Law $E = IR$

A. Set up the circuit as depicted in Figure 5-1.

B. Apply 10 Volts.

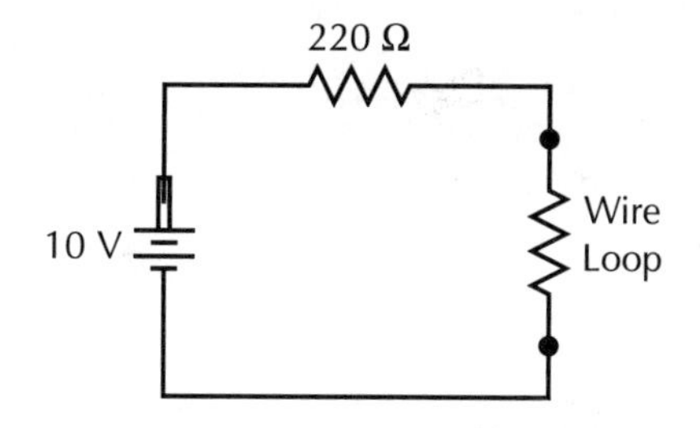

Figure 5-1 Circuit Diagram

C. Measure the current.

D. Measure the voltage across the wire loop.

Current = ____________.

Voltage = ____________.

R = ____________.

8. Measure the resistance with an ohmmeter.

   R = ____________.

9. Fill in Table 5-2.

Measurement Type	Value in Ohms
Resistivity Formula	
Wire Table	
Ohm's Law	
Ohmmeter	

Table 5-2

Questions

1. What is the function of the resistor in Figure 5-1?

____

____

____

____

____

2. How did you determine the gauge?

3. Which method gave the least error?

4. Which method is the best for a technician to use?

5. How does the phone company determine the resistance of its lines?

Experiment 6

Determining Wire Length by Capacitance

Name ______________________ Class __________ Date ____________

Objectives Upon completion of this experiment, you should be able to:

- Determine wire length by measuring with a capacitance meter.
- Determine wire length by measuring with a capacitance bridge.

Text Reference Snyder, *Introduction to Telecommunications Networks*
Chapter 3, Section 1

Materials and Equipment
Audio Frequency Generator
Category-3 Wire, 4-Conductor at a Minimum of 500 Feet
Capacitance Meter
Capacitance Substitution Box (CSB) (2)
Decade Resistance Box (DRB)
Oscilloscope
Resistor, 1000 Ω

Introduction

The measurement of wire length is essential in determining whether a loop is capable of supporting digital technologies such as DSL. Line length can be determined by measuring the capacitance of the line. Line capacitance is fairly standard and can provide an accurate measurement of length. In this experiment you will determine the length of a loop of wire by using a laboratory-constructed bridge as well as a capacitance meter.

Procedures

1. The typical value of telecommunications wire used in local loops is 0.083 μF/mile.

   A. Calculate a capacitance/foot value.

   B. __________ μF/foot.

2. Length by capacitance bridge.

   A. Construct the circuit in Figure 6-1.

   B. To check your circuit, connect a second CSB in place of the unknown capacitor C_x.

   (1) Set the values of the two CSBs equal to each other.

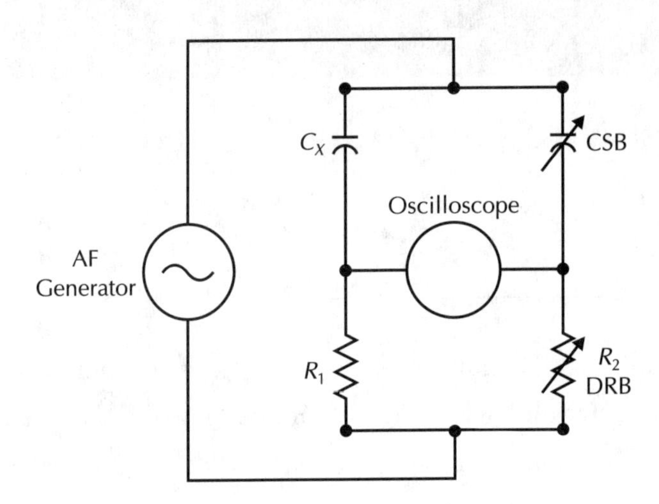

Figure 6-1 Circuit Diagram

(2) Set the DRB to 1000 Ω.

(3) Set the AF generator to 1 kHz.

(4) If the bridge is balanced you should see a minimum voltage condition on the oscilloscope.

Verified: ___________

(5) Unbalance the bridge to confirm the minimum voltage condition.

C. Remove one of the CSBs and replace it with the wire thats length you will be measuring.

D. Balance the bridge by varying the CSB and the DRB for a minimum voltage condition on the oscilloscope.

E. The capacitance of the wire pair is calculated using the following formula:

$$C_x = \frac{C_S \bullet R_2}{R_1}$$

F. The length of the wire is calculated using the following formula:

$$Length\ (\text{ft}) = \frac{C_x}{C_{\text{foot}}}$$

C_{foot} = Capacitance per foot value from step 1B

G. Enter the calculated length in Table 6-1.

Length from step 2	
Length from step 3	
Actual length from instructor	

Table 6-1

3. Length by capacitance meter.

   A. Measure the capacitance of the line using a capacitance meter. See Figure 6-2.

   B. The length of the wire is calculated using the following formula:

$$Length\ (\text{ft}) = \frac{C_x}{C_{\text{foot}}}$$

   C_{foot} = Capacitance per foot value from step 1B

   C. Enter the calculated length in Table 6-1.

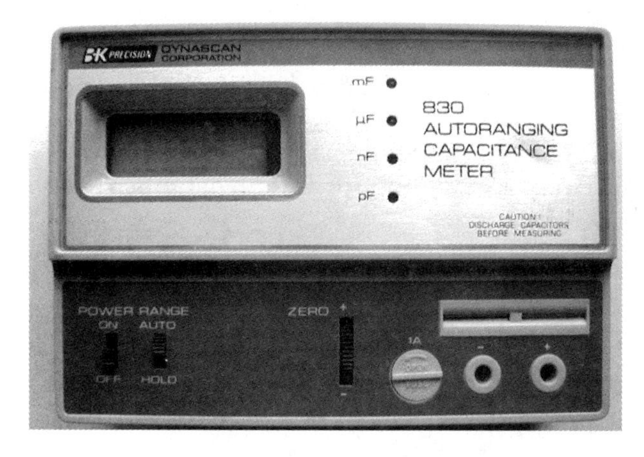

Figure 6-2 Capacitance Meter

Questions

1. How do the length values compare with the actual length?

____

____

____

____

____

2. What error may have been introduced?

____

____

____

____

____

3. What is the maximum loop length for use with DSL?

4. How are legacy POTS loop-length measurements taken?

5. When qualifying a local loop for DSL, what other factors have to be considered besides length?

6. What are the metallic tests that are performed on the local loop?

Experiment 7

Crosstalk

Name ______________________ Class __________ Date __________

Objectives Upon completion of this experiment, you should be able to:

- Demonstrate the effects of crosstalk.
- Measure the effects of crosstalk.
- Differentiate between NEXT and FEXT.

Text Reference Snyder, *Introduction to Telecommunications Networks* Chapter 4, Section 1

Materials and Equipment

Telephone (2)
Category-3 Wire, 4-Conductor
Multimeter
TCOM 100 Trainer
Capacitors, 0.01 μF and 0.1 μF
Punch-Down Tool

Introduction

Crosstalk occurs when electrical noise caused by an electromagnetic field on one conductor is electromagnetically coupled into another conductor in close proximity. Near-end crosstalk (NEXT) occurs between signals in opposite directions. Far-end crosstalk (FEXT) occurs between two adjacent signals transmitted in the same direction. One of the effects of crosstalk in long phone lines is when one person on one phone line hears another person's conversation on another phone line. A capacitance in the wires is created because the wires are too close together, thus allowing AC currents to pass. This current can cause problems in voice and data transmissions. Voice crosstalk is considered an annoyance, but with digital transmission it can cause bit misinterpretation, thus requiring retransmission.

Procedures

1. Remove power from the trainer.
2. Punch down a 9 in. wire from the left side of both the 110 block PB6 blue terminal and the green terminal to the breadboard utilizing two vertical connections. See Figure 7-1.
3. Connect power to the trainer.
4. Plug one phone into J1.

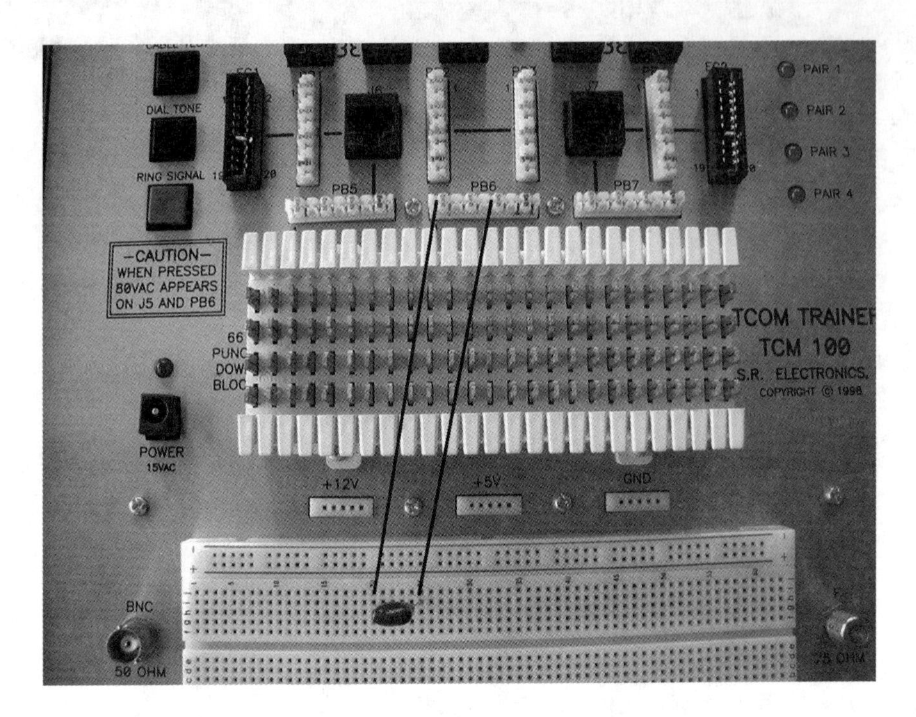

Figure 7-1 PB6 Blue Terminal to Breadboard

5. Plug the other phone into J4.

6. Take both phones off hook.

7. Verify that both phones work by pressing a number button and making sure that you hear a tone in the phone you are pressing the button on.

   Verified: ___________

8. Verify that the phones are not connected by listening on one phone while the other phone's buttons are pressed.

   Verified: ___________

9. Insert a 0.01-μF capacitor in parallel with the two wires on the breadboard, as depicted in Figure 7-1. This will simulate the effect of crosstalk.

10. Perform a calculation that will help you determine what you should expect for an output. Use a frequency of 1000 Hz. What formula did you use and why?

    ____

    ____

    ____

    ____

    ____

11. Verify that the phones are connected to each other by listening on one phone while the other phone's buttons are pressed.

    Verified: ___________

12. Talk into one phone while listening on the other phone.

13. Do you hear voices? Describe.

____

____

____

____

____

14. Insert a 0.1-µF capacitor in parallel with the two wires on the breadboard. See Figure 7-1.

15. Perform a calculation that will help you determine what you should expect for an output. What formula did you use and why?

____

____

____

____

____

16. Verify that the phones are connected to each other by listening on one phone while the other phone's buttons are pressed.

    Verified: ___________

17. Talk into one phone while listening on the other phone.

18. Do you hear voices? Describe.

____

____

____

____

____

19. Draw a schematic of what the circuit looks like.

20. Digital subscriber lines operate at a much higher frequency than voice lines. Calculate the effect of using the same crosstalk capacitance that was used in the experiment to determine the degree of crosstalk interference. DSL operates between 80 kHz and 1.1 MHz.

Questions

1. What are the differences in output between the two capacitors? Do your calculations support your outcome?

2. Based on your calculations, where is crosstalk more of a problem: on voice lines or digital lines? Describe some of the effects of crosstalk.

3. How is crosstalk reduced?

Experiment 8

Coaxial Cable

Name ______________________ Class ____________ Date ______________

Objectives Upon completion of this experiment, you should be able to:

- Identify different types of coaxial connectors.
- Strip wires for termination.
- Crimp an F connector onto a coaxial cable.
- Solder or crimp on a BNC connector.
- Test coaxial cables.
- Distinguish between different types of coaxial cables.

Text Reference Snyder, *Introduction to Telecommunications Networks*
Chapter 4, Section 1

Materials and Equipment
Coaxial Cable RG-58/U
Coaxial Cable RG-59
F Connectors for RG-59 (2)
BNC Connectors (2)
Coaxial Wire Stripper
F-Type Crimping Tool
Soldering Iron
Solder
Multimeter

Introduction

Ernst Werner von Siemens, for whom the unit of conductance (the *siemens*) is named, received a German patent for coaxial cable in 1884. At the time of his death in 1892, no practical use of his invention had yet been made.

Coaxial cable used for broadband transmission was invented by AT&T in 1929 out of a need to handle increased call volume. AT&T established its first cross-continental coaxial transmission system in 1940. Today, coaxial cable is used by cable TV companies between the community antenna and user homes, satellite receivers, over-the-air antenna systems, and in networking.

Coaxial cable is called *coaxial* because it includes an inner core for signal transmission and an outer core that is either a metallic braid or foil and that acts as a signal ground.

Procedures

1. Wire identification.

   A. RG-58 coaxial cable has an impedance of 50 Ω with an outside diameter of 0.18 in. See Figure 8-1.

   B. RG-59 coaxial cable has an impedance of 75 Ω with an outside diameter of 0.14 in.

Figure 8-1 RG-58 Coax

2. Connector identification.

   A. BNC connectors provide a bayonet-locking coupling for quick connect/disconnect. See Figure 8-2.

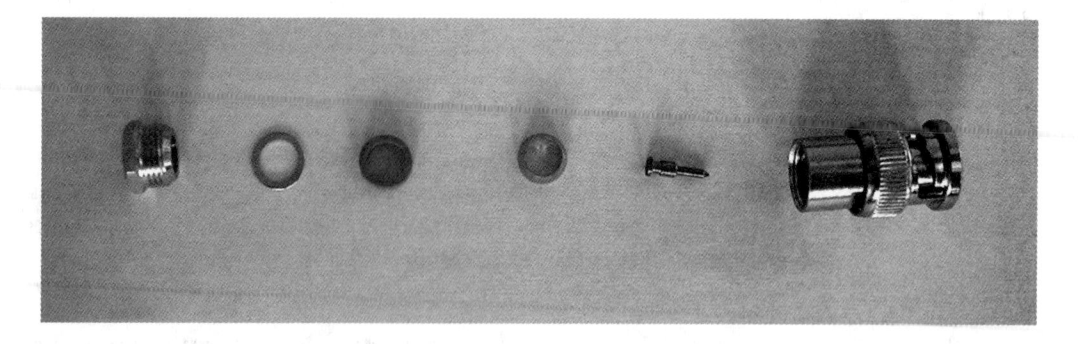

Figure 8-2 BNC Connector

   B. F connectors are used for RF and video connections and are screwed on. See Figure 8-3.

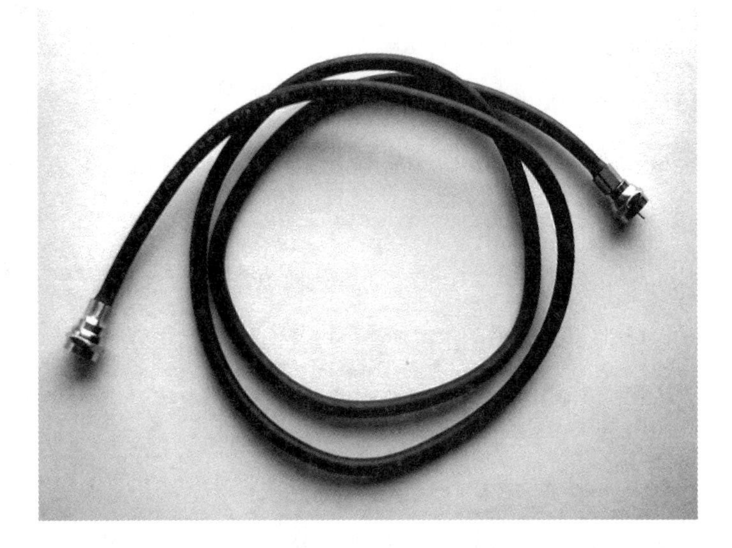

Figure 8-3 F Connector

3. RG-58 termination with solder-type BNC connectors.

   A. Cut a 6-foot length of RG-58 cable.

   B. Follow the procedure below and terminate both ends of the cable.

   C. Place a nut, washer, and gasket over the cable.

   D. Using a coaxial stripper, strip the cable to the proper dimensions. Refer to Figure 8-4.

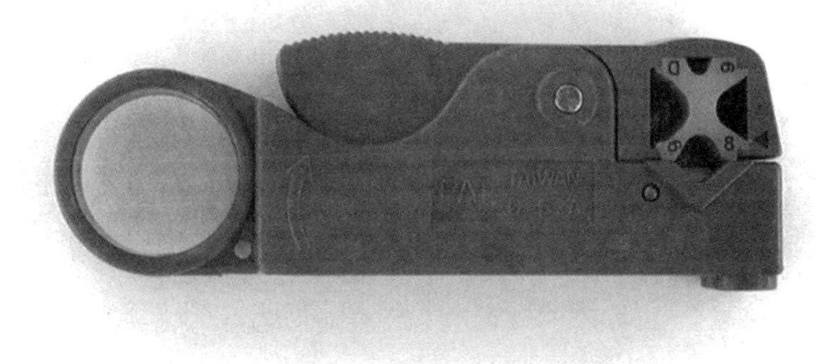

Figure 8-4 Coaxial Stripper

   E. Remove 1/4 in. of the jacket material.

   F. Comb out the braid and fold out.

   G. Trim the insulation on the center conductor by .094 in.

   H. Pull braid wires forward and slide on the clamp.

   I. Fold back the braid wires over the clamp and trim to the proper length.

   J. Solder the male contact to the center conductor.

K. Insert the cable with the male contact into the plug body and tighten to approximately 15 in. pounds.

L. Test the cable construction.

(1) Determine the following expected measurements:

a. Center conductor to outside shell of connector ___________.

b. Center conductor to center conductor ___________.

c. Outside shell to outside shell ___________.

(2) Using an ohmmeter, measure the resistance between the center conductor and the outside of the BNC connector.

Measured resistance ___________.

(3) Using an ohmmeter, measure the resistance between the center conductor to the center conductor of the two BNC connectors.

Measured resistance ___________.

(4) Using an ohmmeter, measure the resistance between the outside shells of the two BNC connectors.

Measured resistance ___________.

4. RG-59 termination with crimp-type F connectors with crimp ring.

A. Cut a 6-foot length of RG-59 cable.

B. Follow the procedure below and terminate both ends of the cable.

C. Using a coaxial stripper, strip the cable to the proper dimensions.

D. Remove 3/8″ of the jacket material. Refer to Figure 8-5.

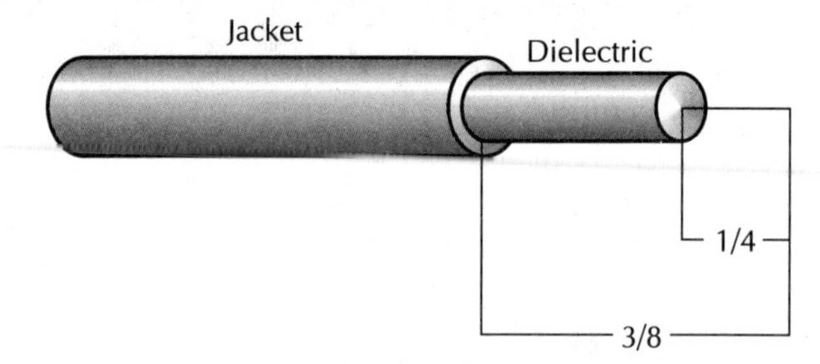

Figure 8-5 F-Cut Dimensions

E. Fold the braid over the jacket.

F. Remove the foil and cut the dielectric length to 1/4 in.

G. Slide the crimp ring over the folded back braid.

H. Rotate and push the connector over the foil and underneath the braid.

I. Slide the crimp ring into place (1/8 in. from the connector body) and crimp with a crimping tool. See Figure 8-6.

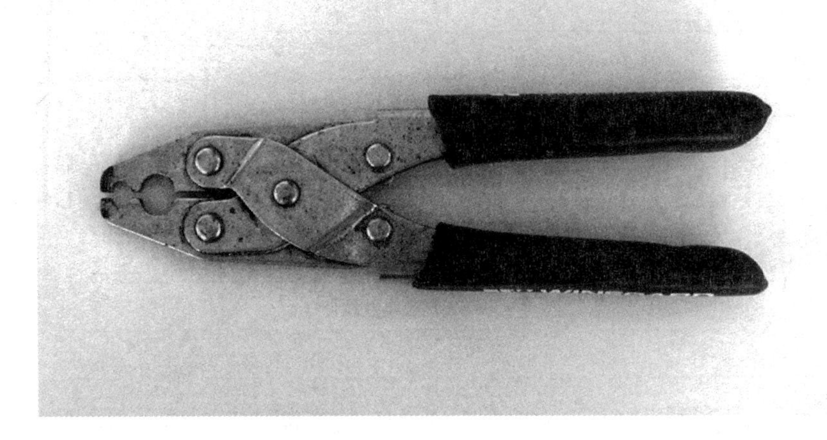

Figure 8-6 F-Type Crimping Tool

J. Test the cable construction.

(1) Determine the following expected measurements:

a. Center conductor to outside shell of connector ___________.

b. Center conductor to center conductor ___________.

c. Outside shell to outside shell ___________.

(2) Using an ohmmeter, measure the resistance between the center conductor and the outside of the F connector.

Measured resistance ___________.

(3) Using an ohmmeter, measure the resistance between the center conductor to the center conductor of the two F connectors.

Measured resistance ___________.

(4) Using an ohmmeter, measure the resistance between the outside shells of the two F connectors.

Measured resistance ___________.

Questions

1. Did your measured values equal your expected measurements?

____

____

____

____

____

2. What does BNC stand for?

3. What does RG stand for in the wire-type designation?

4. What does the /U stand for in the wire designation RG-58/U?

Experiment 9

Fiber-Optic Cable

Name ______________________ Class __________ Date ____________

Objectives Upon completion of this experiment, you should be able to:

- Understand fiber construction.
- Identify tools associated with fiber termination.
- Terminate a fiber connection.
- Understand the necessary precautions when working with fibers.

Text Reference Snyder, *Introduction to Telecommunications Networks*
Chapter 4, Section 3

Materials and Equipment

62.5/125 Jacketed Fiber-Optic Cable, 10 Feet
SC Fiber-Optic Connector Kit (2)
Fiber Cable Stripper
Fiber Stripper
Kevlar Cutting Scissors
5-Minute Epoxy
Fiber Crimp Tool
Cleave Tool
Double-Sided Tape
Microscope
Alcohol Pad
Multi-Grit Fiber Polishing Paper, 5 μm,1 μm, 0.3 μm
Polishing Bushing
Polishing Pad
Polishing Plate

Introduction

Fiber-optic cable can provide higher bandwidths, is relatively immune to EMI, and is lightweight, small, and secure. Fiber cables, like copper cables, have to be terminated. In this experiment you will terminate a 62.5/125-μm fiber cable with multimode SC connectors. Systems consist of a transmitter that converts an electrical signal into light, the optical fiber, and a receiver that converts the light back into an electrical signal.

Procedures

1. The SC connector.

   A. Inspect and verify that you have the following SC-connector components. See Figure 9-1.

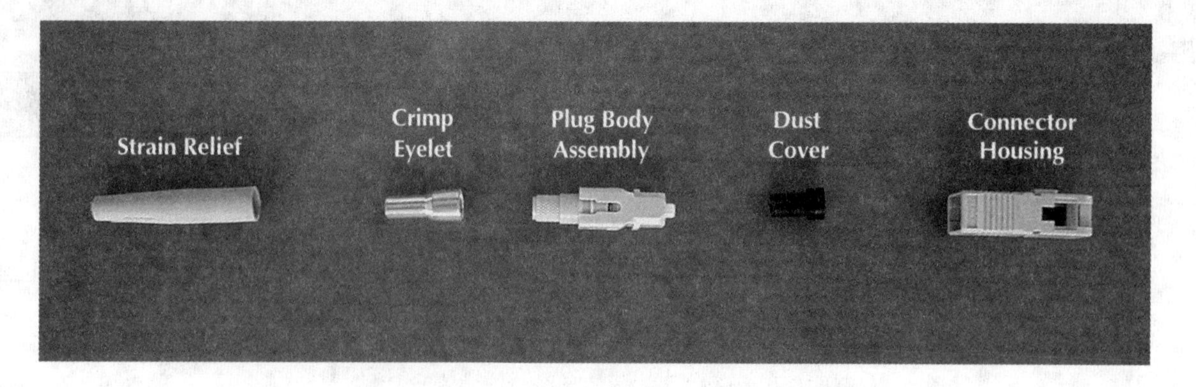

Figure 9-1 SC-Connector Components

(1) Connector Housing ___________

(2) Crimp Eyelet ___________

(3) Dust Cover ___________

(4) Plug Body Assembly ___________

(5) Strain Relief ___________

2. Fiber preparation.

   A. Always wear eye protection.

   B. Always dispose of fiber ends properly, as the ends can create slivers and puncture the skin.

   C. Cut the cable one inch longer than the required length.

   D. Slide the strain relief onto the cable. See Figure 9-2.

   E. Slide the crimp eyelet onto the cable with the wider end toward the front of the cable.

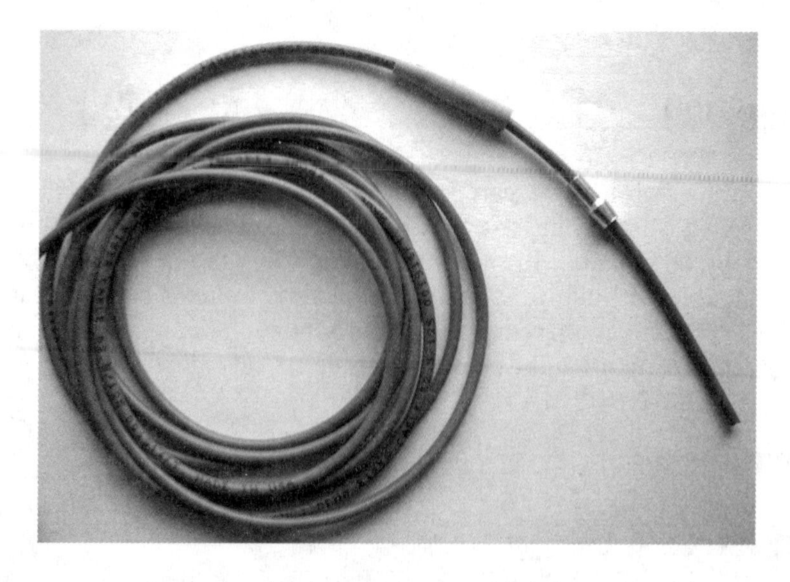

Figure 9-2 Eyelet on Cable

F. Strip the cable to the dimensions shown in Figure 9-3 using fiber-cable strippers and scissors. See Figure 9-4. Dimensions are in inches and are not to scale. Clean the exposed fiber with an alcohol pad. Never clean the fiber with a dry pad.

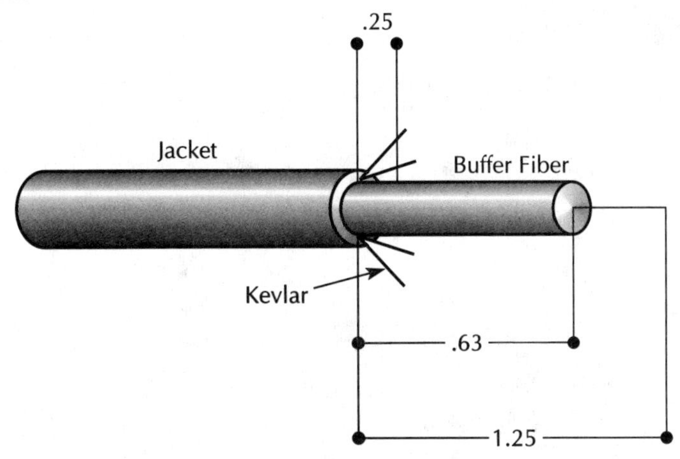

Figure 9-3 Dimensions

Figure 9-4 Fiber-Cable Strippers

G. Slide the plug-body assembly over the fiber. The fiber should extend past the end of the connector and the buffer should stop against the back of the connector.

Verified: ___________

3. Terminating fibers.

A. The epoxy should have an applicator capable of fitting into the end of the plug-body assembly.

B. Mix the epoxy according to the manufacturer's instructions.

C. Insert the epoxy applicator tip into the plug-body assembly and inject the epoxy until epoxy appears at the ceramic tip. Remove the applicator without adding additional epoxy into the connector.

D. Insert the fiber into the connector and flair the strength members over the knurl on the connector.

Verified: ___________

4. Crimping the crimp eyelet.

A. Open the tool fully. See Figure 9-5.

Figure 9-5 Crimper

B. Place the crimp eyelet over the strength members that are over the knurl on the connector.

C. Align the large die in the crimp tool with the edge of the large diameter of the crimp eyelet.

D. Squeeze the crimp at the largest diameter of the eyelet.

E. Align the small die in the crimp tool with the edge of the small diameter of the crimp eyelet.

F. Squeeze the crimp at the smallest diameter of the eyelet.

Verified: ___________

G. Apply a thin layer of epoxy over the crimp eyelet.

H. Slide the strain relief over the entire crimp eyelet.

I. Slide the connector housing over the plug body until it clips into place. Be careful not to break the fiber.

J. Install a fiber protector onto the connector.

K. Position the connector vertically and let the epoxy cure. Times will vary.

5. Cleaving.

A. After the epoxy is cured, remove the fiber protector from the connector.

B. While supporting the connector assembly, use the cleave tool to scribe the fiber. After scribing the fiber, pull the fiber away from the connector and dispose of it properly. See Figure 9-6.

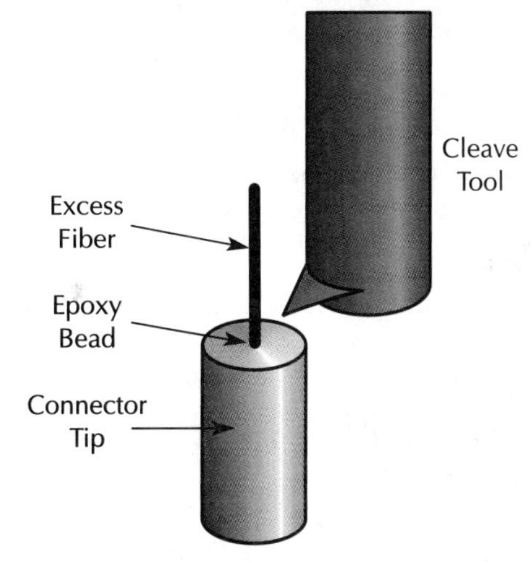

Figure 9-6 Cleave Tool

C. Do not let the cleave tool make contact with the epoxy.

6. Polishing by hand.

A. To level the fiber, lightly polish the end with 5-µm polishing film.

B. Install the connector into the polishing bushing.

C. Place the polishing pad onto the polishing plate.

D. Place the 5-µm polishing film onto the polishing pad.

E. Start polishing in a figure-eight pattern, as shown in Figure 9-7.

F. Stop polishing after completing 20 figure-eight patterns.

G. Clean the tip with the alcohol pad.

H. Remove the 5-µm polishing film.

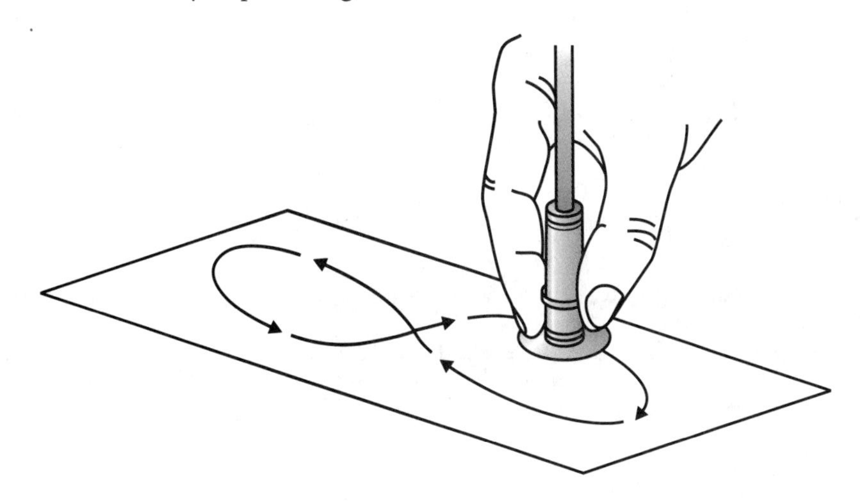

Figure 9-7 Figure-Eight Pattern

I. Place the 1-µm polishing film onto the polishing pad.

J. Start polishing in a figure-eight pattern, as shown in Figure 9-7.

K. Stop polishing after completing 20 figure-eight patterns.

L. Clean the tip with the alcohol pad.

M. Remove the 1-µm polishing film.

N. Place the 0.3-µm polishing film onto the polishing pad.

O. Start polishing in a figure-eight pattern, as shown in Figure 9-7.

P. Stop polishing after the epoxy is removed.

Q. Clean the tip with the alcohol pad.

7. Inspect the fiber.

   A. Be sure all epoxy is removed.

   B. Make sure it is clean.

   C. Inspection is done with a microscope. See Figure 9-8.

Figure 9-8 Microscope

   D. Large scratches and pits are not acceptable.

   E. Small scratches and small chips on the outside of the fiber are acceptable.

Questions

1. What precautions were necessary when terminating fiber connections?

____

____

____

____

____

2. How did your fiber look?

____

____

____

____

____

Experiment 10

ISDN

Name ______________________ Class __________ Date ____________

Objectives Upon completion of this experiment, you should be able to:

- Understand ISDN terms and acronyms.
- Understand ISDN device types.
- Understand the differences between ISDN and traditional modems.

Text Reference Snyder, *Introduction to Telecommunications Networks*
Chapter 6, Section 2

Materials and Equipment Computer with Internet Access

Introduction

Integrated Services Digital Technology (ISDN) originated out of a need for higher bandwidth services for small offices and dial-in users than what was available with traditional dial-in phone lines.

ISDN provides end-to-end digital connectivity and such services as voice, data, and video.

ISDN can also provide backup links in case the main connection is lost. ISDN is digital service. It is a dial-up service that is widely used both domestically and internationally. There are two basic types of ISDN: Basic Rate Interface (BRI) and Primary Rate Interface (PRI).

Procedures

1. Define the following ISDN terms in Table 10-1.

Term	Definition	Device
2B+D		
BRI		
CPE		
DCE		
DDR		
DTE		
LAPD		
NT1		
NT2		
NT12		
PPP		
PRI		
R		
S		
SPID		
SS7		
T		
T1		
T3		
TA		
TE1		
TE2		
U		

Table 10-1

2. Interconnecting ISDN terminal equipment and network termination equipment.

A. Label the interfaces and device types from Figure 10-1.

(1) __________

(2) __________

(3) __________

(4) __________

(5) __________

(6) __________

(7) __________

(8) __________

(9) __________

(10) __________

(11) __________

(12) __________

(13) __________

(14) __________

(15) __________

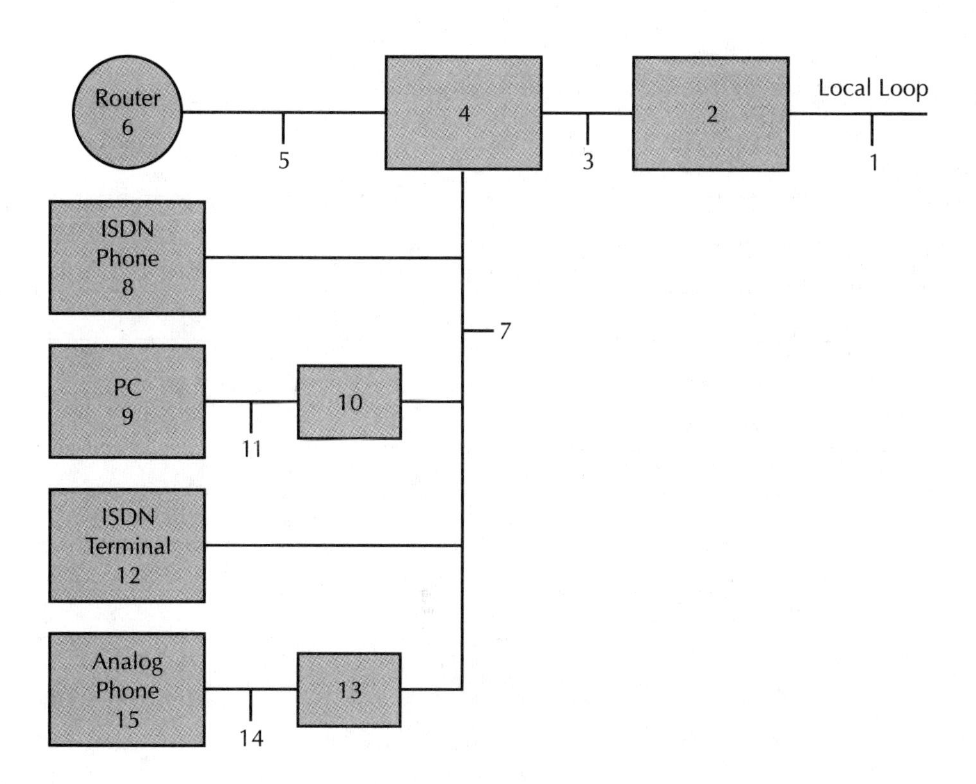

Figure 10-1 ISDN Diagram

Questions

1. What is a SOHO?

2. What is the distance limitation for ISDN? Why?

3. What is the difference between broadband ISDN and narrowband ISDN?

4. What is the function of the Bearer channel?

5. What is the function of the Delta channel?

Experiment 11

RS-232 Serial-Port Testing

Name ______________________ Class __________ Date ______________

Objectives Upon completion of this experiment, you should be able to:

- Understand serial-port terms and abbreviations.
- Make a DB-9 RS-232 loopback tester.
- Test a serial port with an external loopback tester.
- Use a terminal program.

Text Reference Snyder, *Introduction to Telecommunications Networks*
Chapter 7, Section 1

Materials and Equipment
Computer with Internet Access
HyperTerminal Program
DB-9 Female Connector
Category-3 Wire

Introduction

Serial ports are used to connect Data Terminal Equipment (DTE) and Data Communications Equipment (DCE) devices. For example, the DCE is a modem and the DTE is the computer. A serial port is an interface that can be used for interfacing many devices such as modems, mice, and printers. Another example is the connection made from a computer to the console port on a switch or router that allows users to manage the device.

Procedures

1. DB-9 connector.

   A. View the DB-9 female connector from the front and draw the outline diagram of the connector, labeling the pins in the correct order.

B. View the DB-9 female connector from the back and draw the outline diagram of the connector, labeling the pins in the correct order.

C. Fill in Table 11-1 with the pin number associated with the function.

Abbreviation	Function	Pin Number
CTS	Clear to Send	8
DCD		
DSR		
DTR		
Gnd		
RI		
RTS		
RxD		
TxD		

Table 11-1

2. Making the loopback tester.

A. Short out pins 2 and 3 on the female connector, as shown in Figure 11-1.

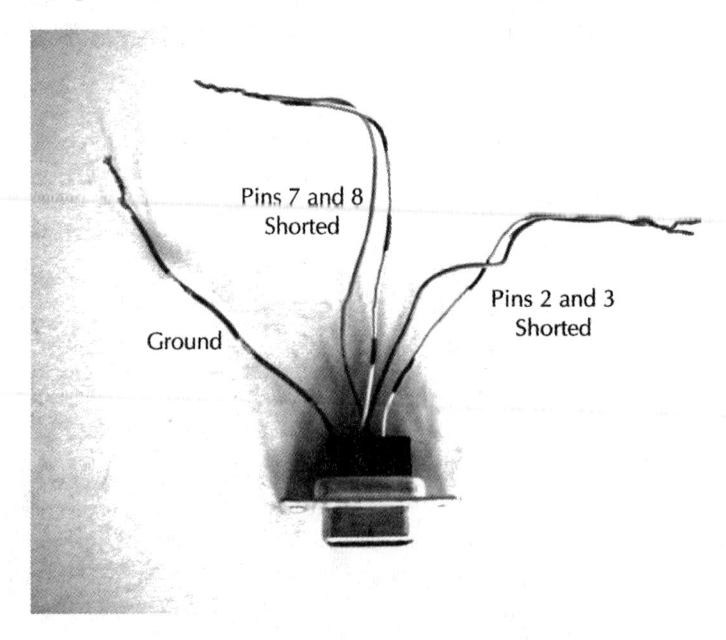

Figure 11-1 Loopback Tester with Pins Shorted

B. Short out pins 7 and 8 on the female connector, as shown in Figure 11-1.

C. Have another person inspect the connector for proper wiring.

Verified: ___________

3. Test the serial port using HyperTerminal *without* the loopback tester.

A. Launch the HyperTerminal program and name the connection "No Loopback," as shown in Figure 11-2, then click **OK**.

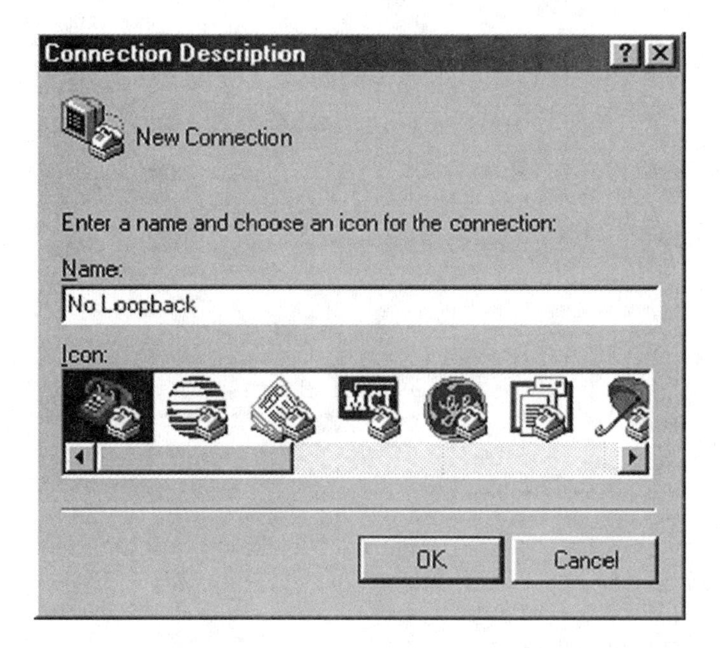

Figure 11-2 HyperTerminal Launch

B. Connect using the direct-to-com port, as shown in Figure 11-3.

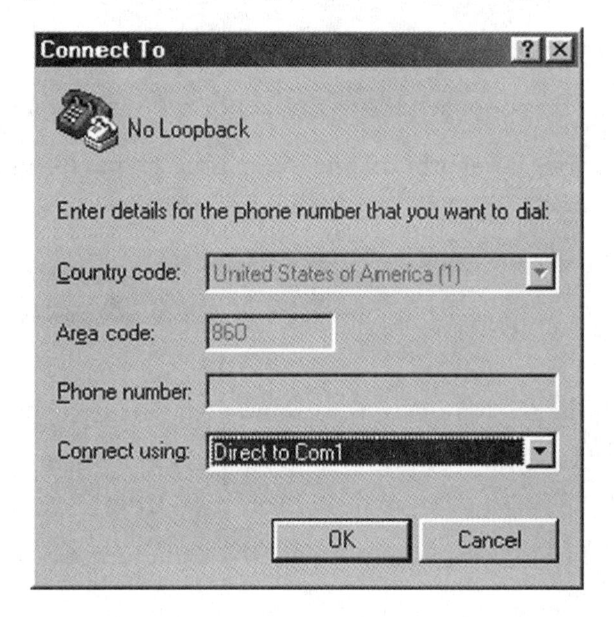

Figure 11-3 Direct-to-Com Port

C. Set the properties as shown in Figure 11-4 and then click **OK**.

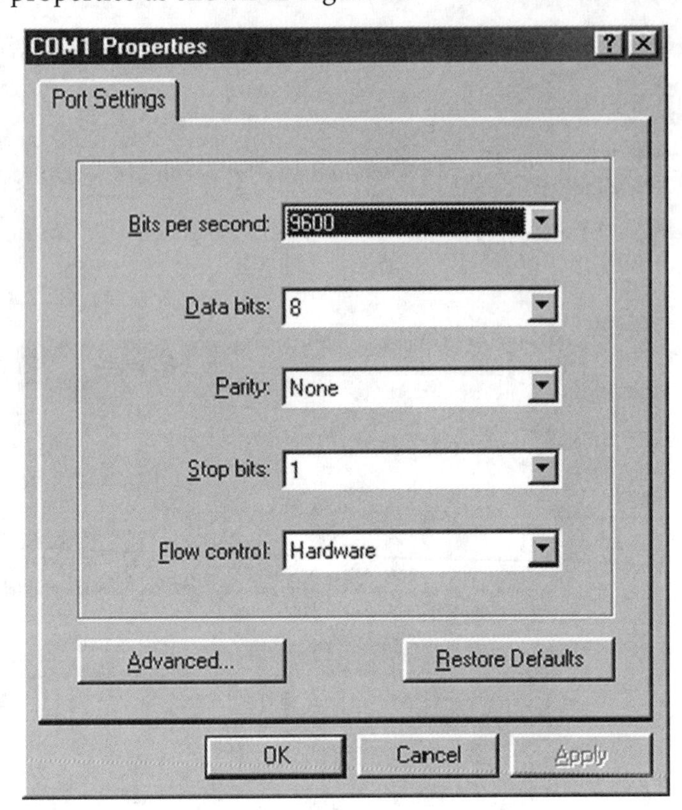

Figure 11-4 Com Properties

D. With the session open, type some text. The text should not be echoed back onto the screen.

Verified: ___________

4. Test the serial port using HyperTerminal *with* the loopback tester.

A. Turn your computer off.

B. Install the loopback tester to the port you will be testing.

C. Turn your computer on and launch the HyperTerminal program and name the connection "Loopback," then click **OK**.

D. Connect to the com port you are using.

E. Set the properties the same as you did in section 3C, Figure 11-4 and then click **OK**.

F. With the session open, type some text. The text should be echoed back onto the screen.

Verified: ___________

G. If you have other com ports on the computer, test them also.

Questions

1. What output would you expect from testing a com port that is being used by an internal modem using HyperTerminal to test the port?

2. What type of device is an external modem: a DTE or a DCE?

3. What allows us to connect two DTEs to each other, computer to computer?

Experiment 12

Analog Modems

Name ______________________ Class __________ Date ______________

Objectives Upon completion of this experiment, you should be able to:

- Install a modem.
- Configure the software in accordance with the user's manual.
- Utilize modem commands.
- Use a terminal program.
- Transfer files via a modem.

Text Reference Snyder, *Introduction to Telecommunications Networks*
Chapter 7, Section 1

Materials and Equipment
Computers with Windows™ Operating System (2)
Analog Modems (2)
TCOM 100 Trainer
Multimeter

Introduction

Analog modems are used in a home or business connected to the local loop to transmit and receive data. We will explore another method for computers to communicate without the local loop through the use of modems and a standard terminal program. The local loop needs to be replaced by a line simulator that can supply at least 14 V and 20 mA DC.

Procedures

1. Fill in Table 12-1 with the function of the modem command. Commands can be found in the owner's manual for the modem or by doing a search on the Internet.
2. You will need two computers for this lab exercise.
3. Install a modem into a computer following the installation instructions.
4. Select a computer to be Computer 1; the other will be Computer 2.
5. Connect each computer modem to J4 on the TCOM 100 trainer.
6. Connect the two J5s on the TCOM 100 trainers together with a patch cable. See Figure 12-1.

Command	Function
AT	
ATI3	
ATX3	
ATA	
ATD	
ATL	
ATE0	
ATE1	

Table 12-1

Figure 12-1 Jack J5 on Trainers

7. Power up the trainers.
8. Measure the voltage of the line simulator at the PB6 brown terminal.

   Voltage = ___________
9. On both computers launch the HyperTerminal program located in the **Windows\ Program Files\Accessories\HyperTerminal** subdirectory. See Figure 12-2.

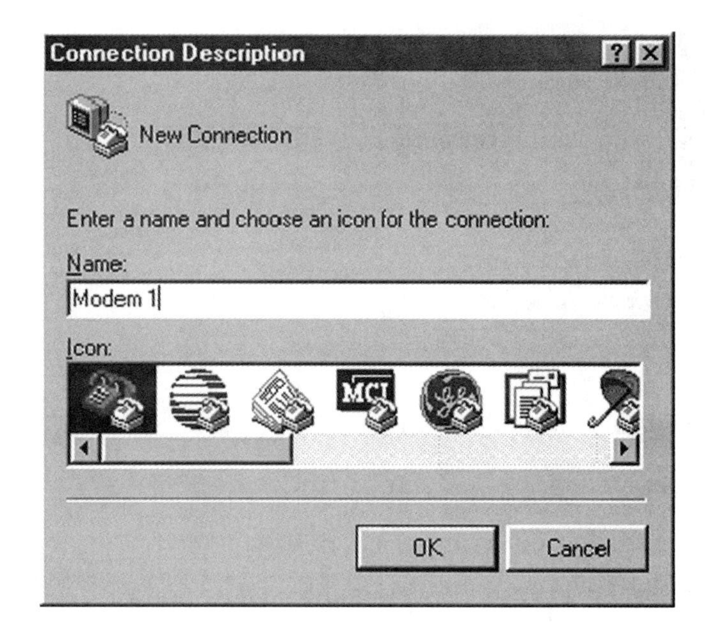

Figure 12-2 HyperTerminal Launch

10. Name the new connection *Modem 1* on Computer 1 and *Modem 2* on Computer 2.

11. Connect using the modem, as shown in Figure 12-3, and click **OK.**

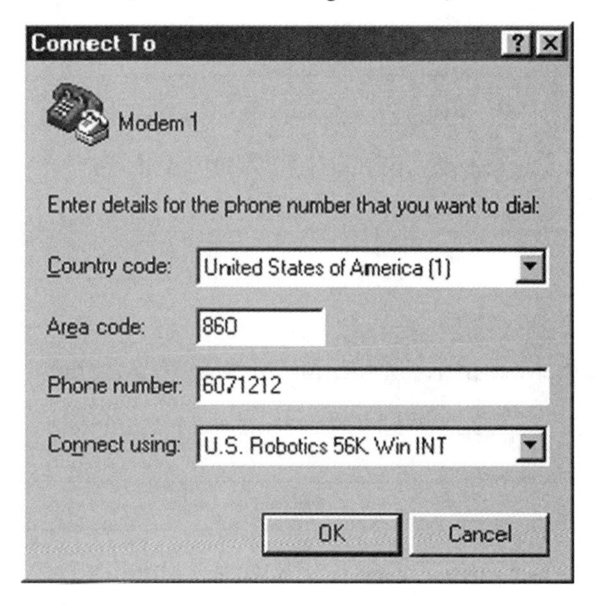

Figure 12-3 Connect Using a Modem

12. Using the terminal program on Computer 1, type in the following commands in the terminal screen and then hit enter:

    A. AT (you should get back the "OK" response)

       Verified: ___________

    B. ATE1

    C. ATX3&C0

    D. ATD

13. Using the terminal program on Computer 2, type in the following commands in the terminal screen and then hit enter:
    A. AT (you should get back the "OK" response)
       Verified: ___________
    B. ATE1
    C. ATX3&C0
    D. ATA
14. The computers should connect.
15. Transfer a file from Computer 1 to Computer 2.
16. Transfer a file from Computer 2 to Computer 1.

Questions

1. What modem command can be issued to display a modem's configuration profile?

____

____

____

____

2. What is the upstream bit rate of a V.90 modem?

____

____

____

3. What is the downstream bit rate of a V.90 modem?

____

____

____

4. How do you find out how your modem handles transmission errors?

____

____

____

____

5. What method does your modem use to detect transmission errors?

____

____

____

____

Experiment 13

Short Haul Modems

Name ______________________ Class __________ Date ____________

Objectives Upon completion of this experiment, you should be able to:

- Understand how to install a short haul modem.
- Understand how to use and set values on a communications port.
- Construct a communications circuit utilizing short haul modems.
- Demonstrate the functionality of a communications circuit using short haul modems.
- Communicate over a circuit using short haul modems.

Text Reference Snyder, *Introduction to Telecommunications Networks*
Chapter 7, Section 1

Materials and Equipment
Computers with a Free Serial Port (2)
Short Haul Modem (2)
Assorted Serial Cables, Straight Through
Category-3 Wire, 4-Conductor, 500 Feet
HyperTerminal Program

Introduction

Short haul modems are devices that are used for extending the distance of electrical signals such as RS-232, V.35, T1, T3, and E1 over twisted pair, coaxial, or fiber cable. Short haul modems can extend the distance up to 5 km and 2048 kbps using standard telecommunications wire. If there are two computers located more than 1000 feet apart that would be out of the distance limitations supplied by most standard interfaces, a short haul modem can be used to allow these devices to communicate as if they were 10 feet apart.

Procedures

1. Attach a short haul modem to a free serial port on each computer following the manufacturer's instructions. See Figure 13-1 for an example.
2. Apply power to the short haul modem.
3. Set the short haul modem into the loopback mode to test the functionality of your installation.

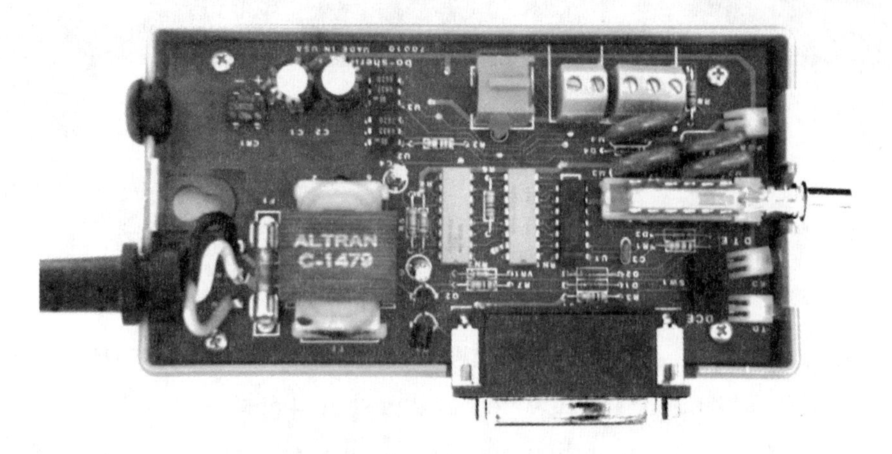

Figure 13-1 Short Haul Modem

4. Start the HyperTerminal Program located in the **Windows\Program Files\Accessories\HyperTerminal** subdirectory and name the connection *Short Haul 1* on Computer 1 and *Short Haul 2* on Computer 2. See Figure 13-2.

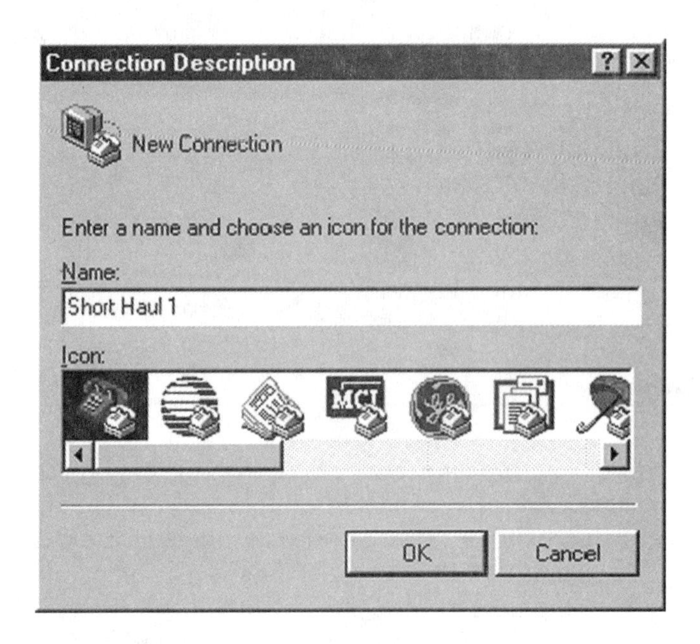

Figure 13-2 HyperTerminal Launch

5. Connect to the Com port you are using, as shown in Figure 13-3.

6. Make sure the port settings on each computer are the same as those depicted in Figure 13-4.

7. From the terminal screen, type some characters and hit enter. Watch the LEDs on the short haul modem—both the TD (transmit data) and the RD (receive data) LEDs should be blinking.

   Verified: ___________

8. You should also see the characters you typed appear on the screen.

   Verified: ___________

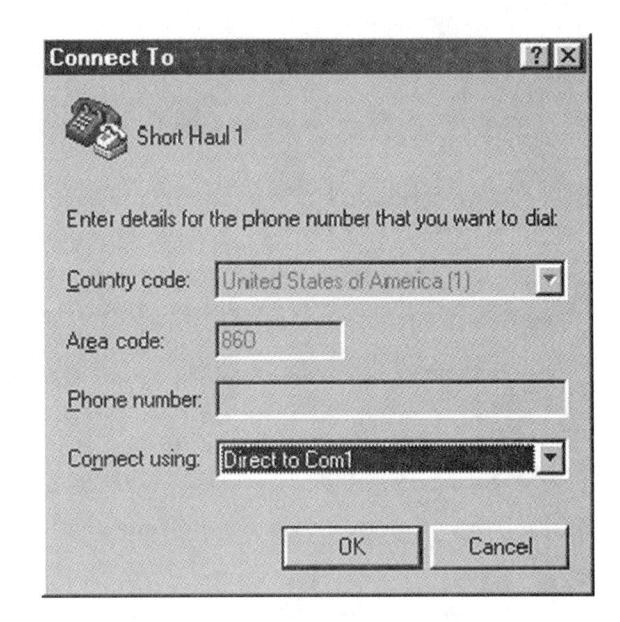

Figure 13-3 Com Port

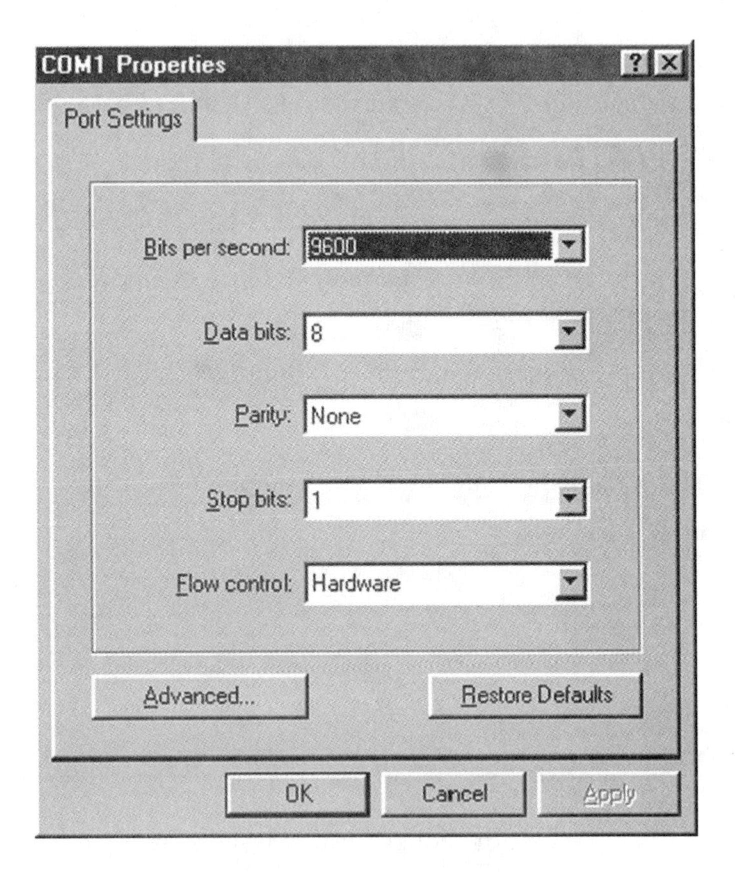

Figure 13-4 Port Settings

9. If this does not work, you must troubleshoot your installation.
10. Test both short haul modems for proper operation.
11. Remove power to the modems.

12. Once the test is completed and working you can connect the short haul modems together.

13. Wire the modems together using the 500-foot coil of CAT-3 wire.

14. Connect the T+ from modem 1 to the R+ on modem 2.

    Wire Color __________

15. Connect the T– from modem 1 to the R– on modem 2.

    Wire Color __________

16. Connect the T+ from modem 2 to the R+ on modem 1.

    Wire Color __________

17. Connect the T– from modem 2 to the R– on modem 1.

    Wire Color __________

18. Set the modems back to their normal operating modes from the loopback mode.

19. Power up the modems.

20. Type some characters on Computer 1 and hit enter. Watch the LEDs on the short haul modem. The TD (transmit data) LED should be blinking.

21. The RD LED on modem 2 should also be blinking.

22. You should be able to read the characters on Computer 2.

23. Repeat steps 18 to 20 on Computer 2. Once this is done you should be able to read the characters on Computer 1.

24. Transfer a file from Computer 1 to Computer 2.

    Verified: __________

26. Transfer a file from Computer 2 to Computer 1.

    Verified: __________

Questions

1. Name three applications for the use of a short haul modem.

____

____

____

____

____

2. What is another name for a short haul modem?

3. Look up the price for a pair of short haul modems.

4. Investigate the costs of performing a task with a short haul modem and with an alternative method of your choice. Compare and contrast the two methods.

Experiment 14

ASCII

Name ______________________ Class __________ Date ____________

Objectives Upon completion of this experiment, you should be able to:

- Understand digital representation of an ASCII character.
- Understand digital representation of a parity bit.
- Understand asynchronous serial transmission.
- Use an oscilloscope to view the digital representation of an ASCII character.

Text Reference Snyder, *Introduction to Telecommunications Networks*
Chapter 7, Section 7

Materials and Equipment
Computer with a Free Serial Port
Oscilloscope Storage
DB-9 Female Connector
HyperTerminal Program
ASCII Table

Introduction

The electronic data communications serial port RS-232 interface was introduced in 1962 and is still widely used in the computer industry. RS-232 data is bipolar with +3 to +12 V indicating a logic 0 and a –3 to –12 V indicating a logic 1.

Asynchronous serial transmission is used by personal computers to connect to printers, modems, and fax machines. ASCII characters are transmitted in a series of bytes along a single wire with a ground. The byte is serialized and sent as a series of bits. In this experiment we will analyze the data bit stream when a keyboard key is pressed by viewing the output of a serial port on a storage oscilloscope.

Procedures

1. Using the loopback tester made in Experiment 11, attach the tester to an open serial port on your computer.
2. In order to see the output on the HyperTerminal screen, leave the loopback tester pins 2 and 3 together so that you can attach an oscilloscope probe to them.
3. Connect the oscilloscope probe to pins 2 and 3 and the ground to pin 5.

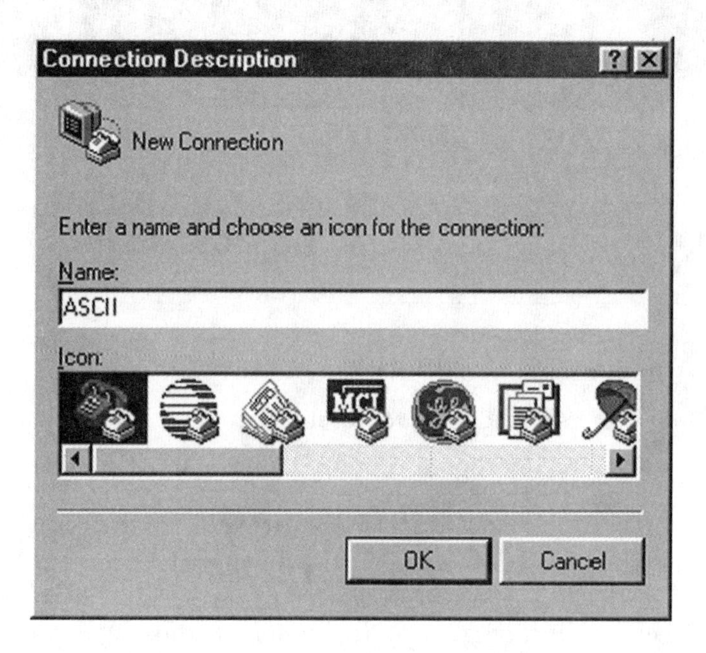

Figure 14-1 HyperTerminal Launch

4. Launch the HyperTerminal program and name the connection ASCII, as shown in Figure 14-1. Click **OK**.
5. Connect using the Direct-to-Com port, as shown in Figure 14-2.

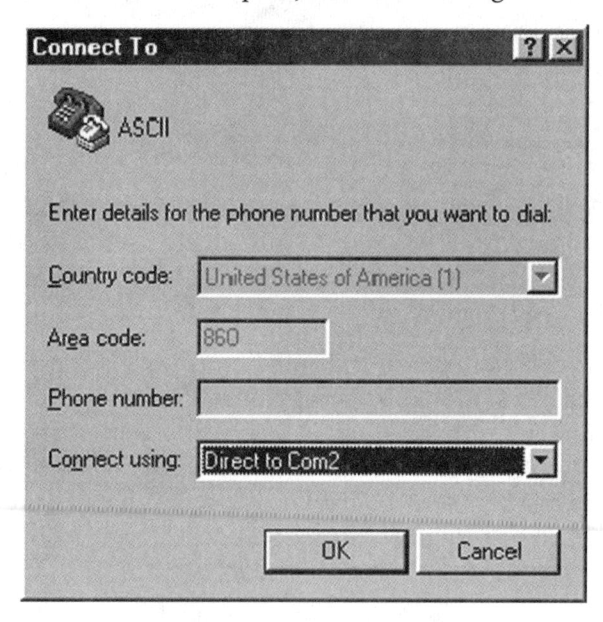

Figure 14-2 Com Port

6. Set the properties as shown in Figure 14-3 and then click **OK**.
7. With the session open, type some text. The text should be echoed back onto the screen.

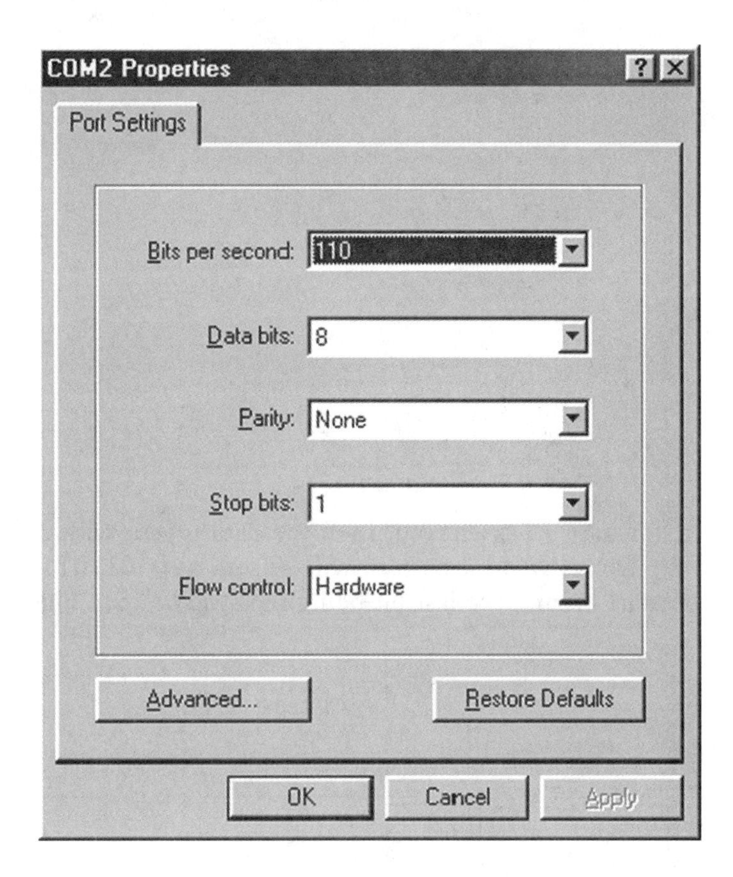

Figure 14-3 Properties

8. Oscilloscope setup.

   A. Determine the sweep speed that will be necessary to view a 110-bps data rate. What we are looking for is a time-base value per cm that will display one bit.

   B. Use the formula (frequency = 1/time).

   C. Result ___________.

9. Set up the oscilloscope as follows:

   A. Channel 1

   B. Amplitude = 5 V/cm

   C. Sweep Speed = Value from 8C

   D. Trigger = Single Sweep

10. From the ASCII table, find the binary value of the letter "m."

    A. Result ___________.

B. Draw the resulting waveform.

C. First a start bit is sent out, then the data is sent with the least-significant bit sent first. The viewed bit pattern will be sent as 0101101101, with the first bit being the start bit and the last bit being the stop bit. See Figure 14-4.

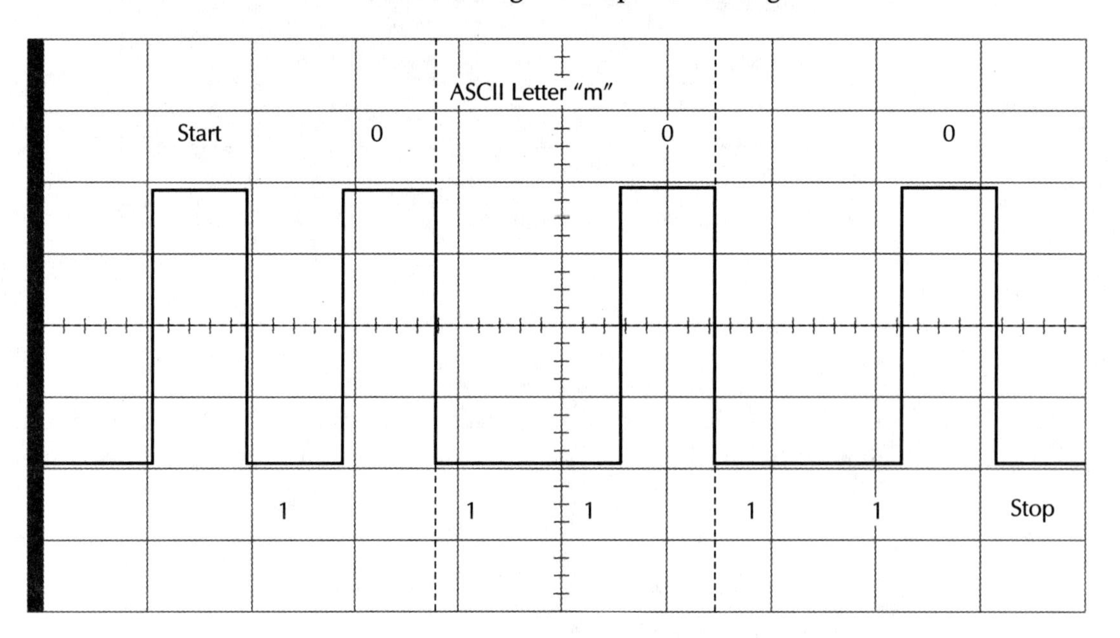

Figure 14-4 Waveform

D. Draw the resulting waveform.

11. View the waveform on the oscilloscope.

    A. Connect the oscilloscope probe to pin 3 on the loopback tester and the ground lead to pin 5.

    B. From the HyperTerminal screen, type the letter "m."

    C. Draw the waveform. (*Hint*: The waveform should look like the one in Figure 14-4 *and* like the waveform you drew in step 10D.)

    D. Have another person in the class type a letter on your keyboard and try to figure out which letter was pressed by analyzing the waveform.

Questions

1. Name three applications for the use of a short haul modem.

2. What is another name for a short haul modem?

3. Look up the price for a pair of short haul modems.

4. Investigate the costs of performing a task with a short haul modem and an alternative method of your choice. Compare and contrast the two methods.

5. What would your sweep speed have to be set at for a 9600-bps data rate?

Experiment 15

Networking Standards

Name ______________________ Class __________ Date ____________

Objectives Upon completion of this experiment, you should be able to:

- Name the seven OSI layers in order.
- Describe the functions of each layer.
- Identify at which layer the physical components operate.
- Describe several protocols that operate at each layer.
- Identify the main features of the IEEE 802 standards.
- Identify a MAC address.

Text Reference Snyder, *Introduction to Telecommunications Networks*
Chapter 9, Section 4

Materials and Equipment Computer with Internet Access

Introduction

Networking standards allow different vendors to produce products that will work together. The OSI model was developed by the International Standards Organization to help provide a common framework for the development of LANS and WANS. The OSI model helps us troubleshoot network problems by breaking down functions into distinct layers. Understanding the OSI model is essential to understanding how a network functions. The IEEE 802 project pertains to LAN standards in the bottom two layers of the OSI model.

Procedures

1. Fill in the blanks relating to the OSI layers in Tables 15-1, 15-2, and 15-3.
2. Fill in the blanks relating to the IEEE 802 standards in Table 15-4.
3. We will investigate three ways to determine the MAC address of your NIC. MAC is a hardware address and is how you computer is identified on the network. MAC addresses are 48 bits in length and are represented in hex. The first 6 hex digits represent the OUI and the remaining 6 digits represent the serial number.

Layer Number	Name	Mnemonic	Layer Function
7	Application	All	
6	Presentation	People	
5	Session	Seem	
4	Transport	To	
3	Network	Need	
2	Data-Link	Data	
1	Physical	Processing	

Table 15-1

Layer Number	Name	Mnemonic	Protocols at this Layer
7	Application	All	
6	Presentation	People	
5	Session	Seem	
4	Transport	To	
3	Network	Need	
2	Data-Link	Data	
1	Physical	Processing	

Table 15-2

Layer Number	Name	Encapsulation Unit	Devices that Operate at this Level
7	Application		
6	Presentation		
5	Session		
4	Transport		
3	Network		
2	Data-Link		
1	Physical		

Table 15-3

Standard	Description
802.1	Sets internetworking standards related to network management
802.2	
802.3	
802.4	
802.5	
802.6	
802.7	
802.8	
802.9	
802.10	
802.11	
802.12	
802.13	
802.14	
802.15	
802.16	

Table 15-4

A. If you are using a Windows 98 OS, then click **Start**, then **Run**, and then type **winipcfg** in the dialog box and click **OK**. The IP Configuration box will open. Make sure the NIC is in the display, as shown in Figure 15-1, and enter in the Adapter Address.

   Adapter Address ___________

B. If you are using a Windows NT OS, then open a DOS prompt and type **ipconfig/all** and hit enter, as shown in Figure 15-2. This method works with all WIN versions. The physical address displayed should match the adapter address obtained from step A.

   Verified: ___________

C. If you are on a network you can use the ping command in DOS. This allows us to check the connectivity between two computers. If you ping your computer from another computer it will generate an entry into its arp table. Arp is used to associate the relationship between the IP address and the MAC addresss.

   (1) From another computer on your network open a DOS window.

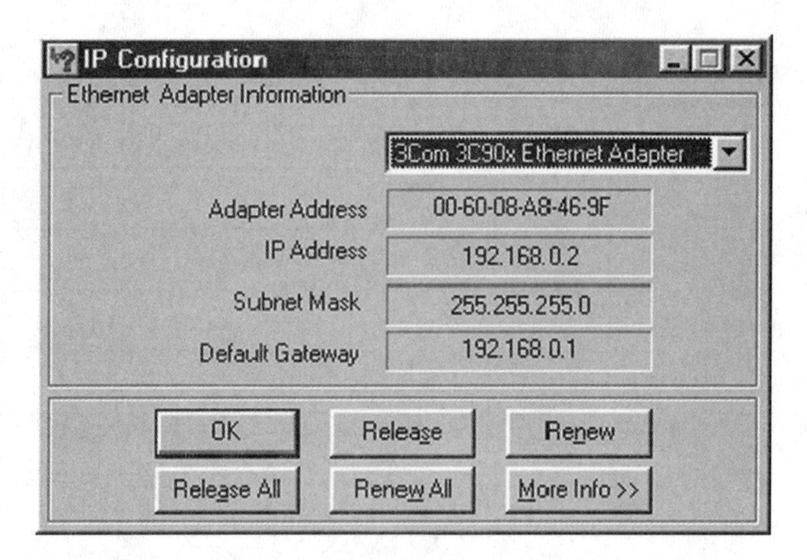

Figure 15-1 Winipcfg

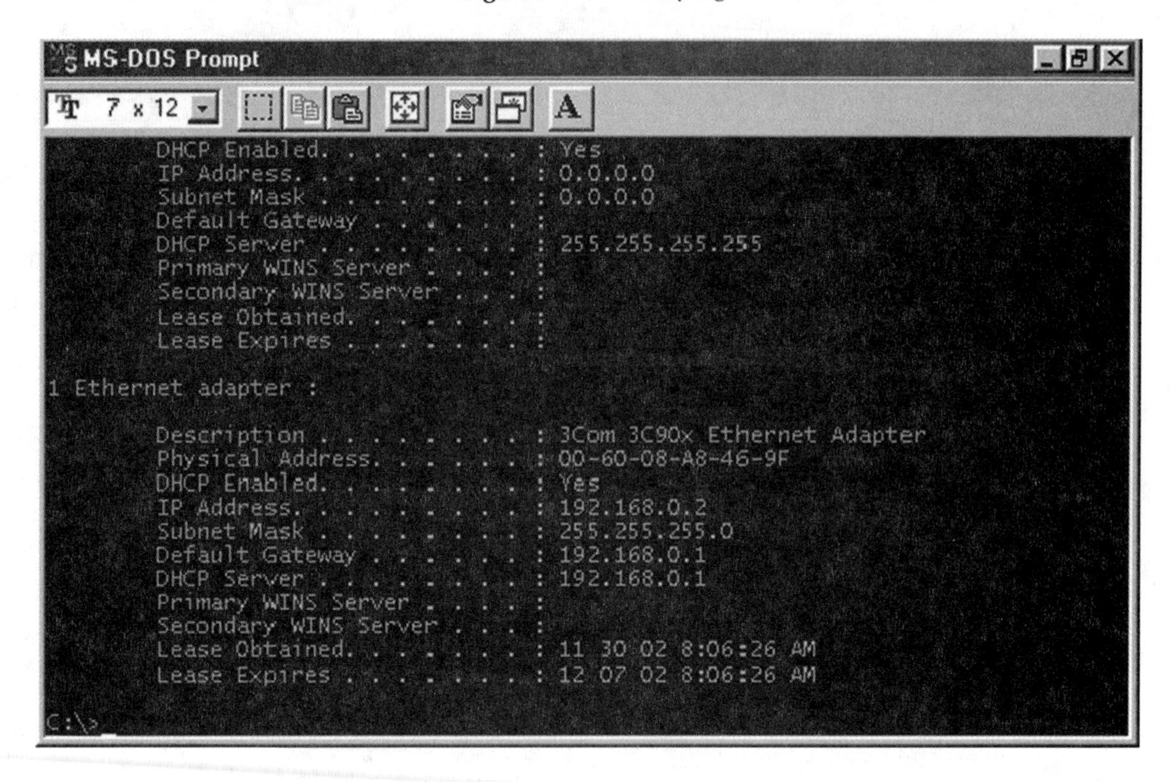

Figure 15-2 Iconfig

(2) Type **ping** (your IP address). I.e. type **C:\>ping 192.168.0.1** and hit **enter**. See Figure 15-3.

(3) You should get a reply back from your computer, as shown in Figure 15-3.

Verified: ____________

(4) Type **arp -g** at the command prompt and then hit **enter** to view the contents of the arp table. Again, see Figure 15-3.

(5) The arp table entry for your IP address should agree with the physical address of your adapter.

Verified: ____________

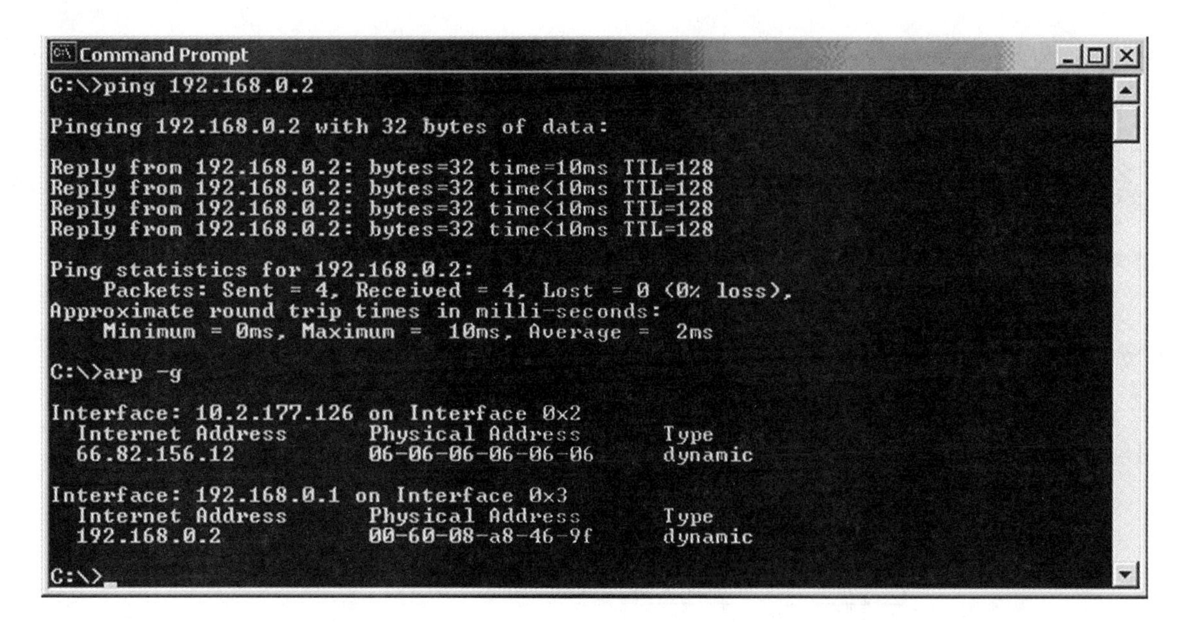

Figure 15-3 Arp

Questions

1. What is the length in bits of a MAC address?

2. What is the length in hex of a MAC address?

3. Which part of the MAC address is the OUI?

4. What is OUI?

5. Given an OUI of 00-60-08, who is the manufacturer?

6. Given an FCC number of EJMNPDSPD035, who is the manufacturer and what is the model of this NIC?

Experiment 16

Peer-to-Peer Networks

Name ______________________ Class ____________ Date ____________

Objectives Upon completion of this experiment, you should be able to:

- Create a peer-to-peer network.
- Understand the role of protocols.
- Understand file sharing.
- Understand access types.
- Understand the use of passwords in a peer-to-peer environment.

Text Reference Snyder, *Introduction to Telecommunications Networks*
Chapter 9, Section 2

Materials and Equipment
Computer with Windows 98 (2)
NIC (2)
Category-5 Patch Cables
Ethernet Switch or Hub
Crossover Cable

Introduction

Peer-to-peer computer networks are computer networks that allow all users on the network to share resources. Resources can include such things as printers, CD-ROM readers, fax machines, scanners, and the contents of hard drives (files). Sharing resources saves time and the expense of duplicating equipment and transferring data. Because security needs to be set up on each computer with passwords given to each user (which can become cumbersome if many resources are shared), the recommended size for peer-to-peer networks is limited to 10 users or fewer.

Procedures

1. You should start with a computer with a properly installed network interface card and a connection to another computer through a switch, hub, or a crossover cable.

2. Protocol installation—*to be done on each computer.*

   A. Install NetBEUI, a small non-routable protocol.

   B. Click **Start/Settings/Control Panel/Network.** Open network by double clicking on the icon.

C. You should see a Network screen similar to that depicted in Figure 16-1.

D. Click **Add.**

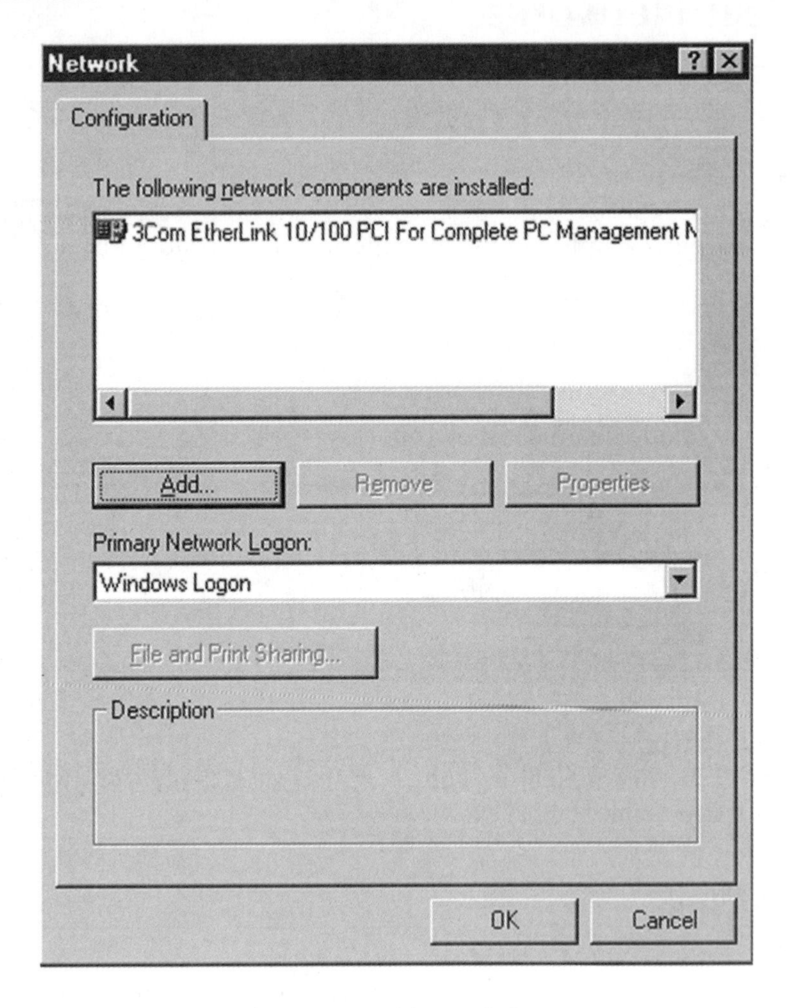

Figure 16-1 Network Screen

E. Click **Protocol** and then **Add.** See Figure 16-2.

F. Select **Microsoft**, then **NetBEUI**, and then click **OK.** See Figure 16-3.

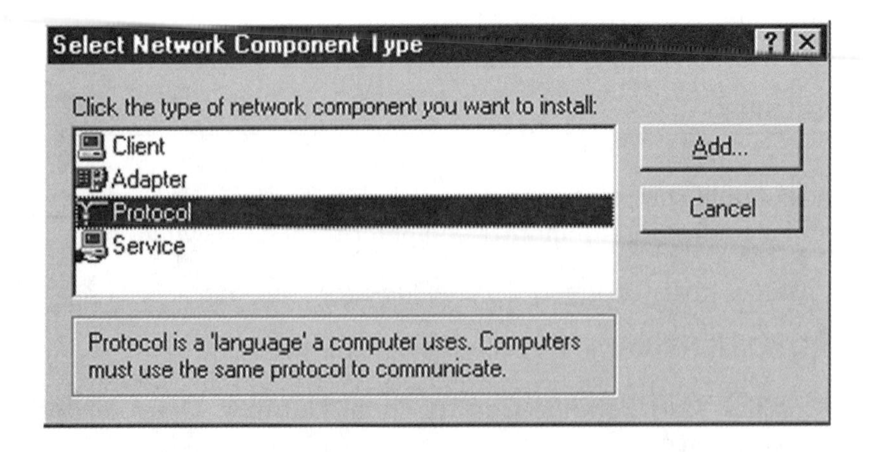

Figure 16-2 Protocol Screen

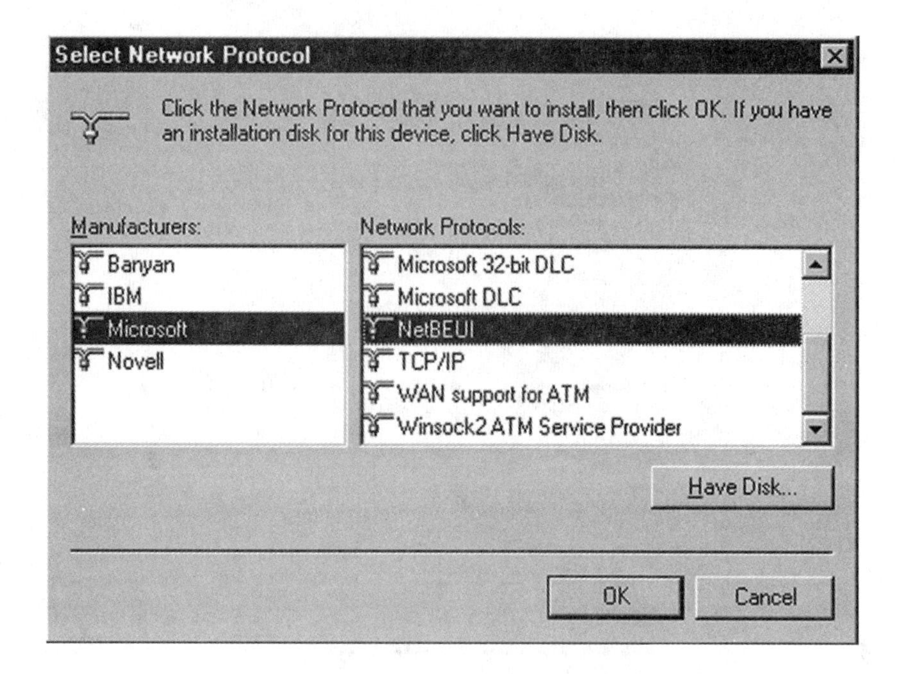

Figure 16-3 NetBEUI

G. Because we will be sharing files, under **File and Print Sharing**, select **I want to be able to give others access to my files.** Then click **OK.** See Figure 16-4.

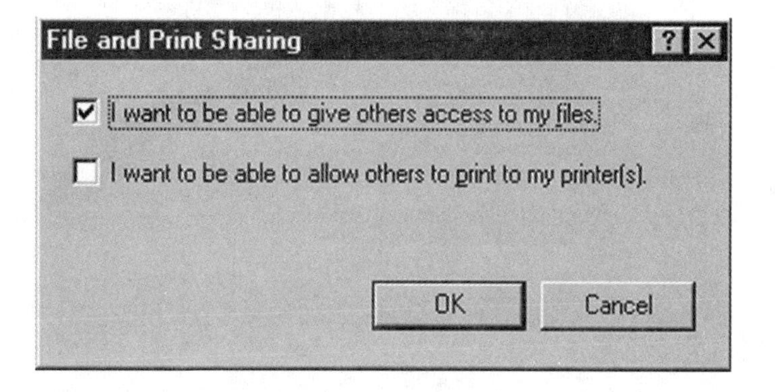

Figure 16-4 Access to Files

H. You will need to know the computer name and be in a workgroup with similar names. **Name** and **Workgroup** can be obtained from the **Identification** tab in the **Network** screen. See Figure 16-5.

Computer 1 name ___________

Computer 2 name ___________

Workgroup name ___________

I. Click the **Access Control** tab. Make sure the radio button for **Share-level access control** is selected. See Figure 16-6.

J. Click the **Configuration** tab and check that the settings match Figure 16-7.

K. Click **OK** to save the Network configuration. Your computer will reboot.

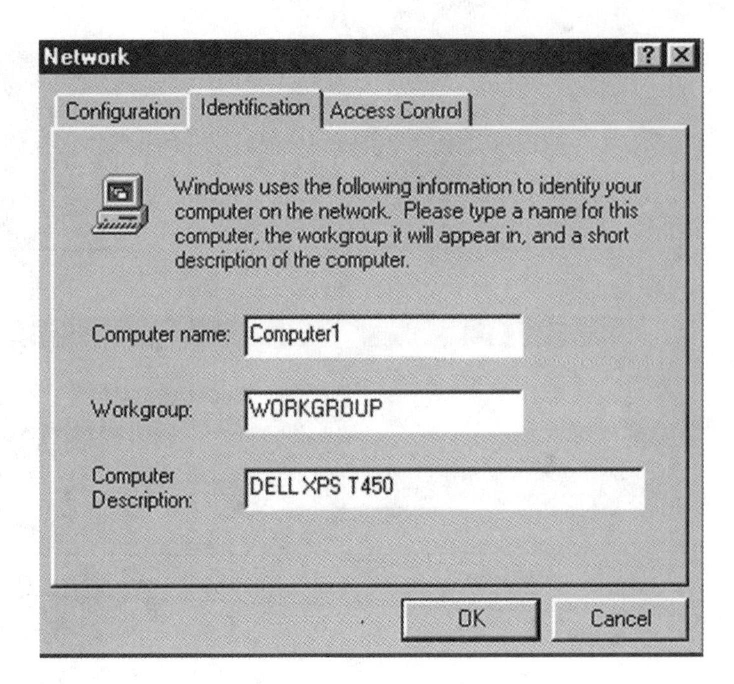

Figure 16-5 Workgroup

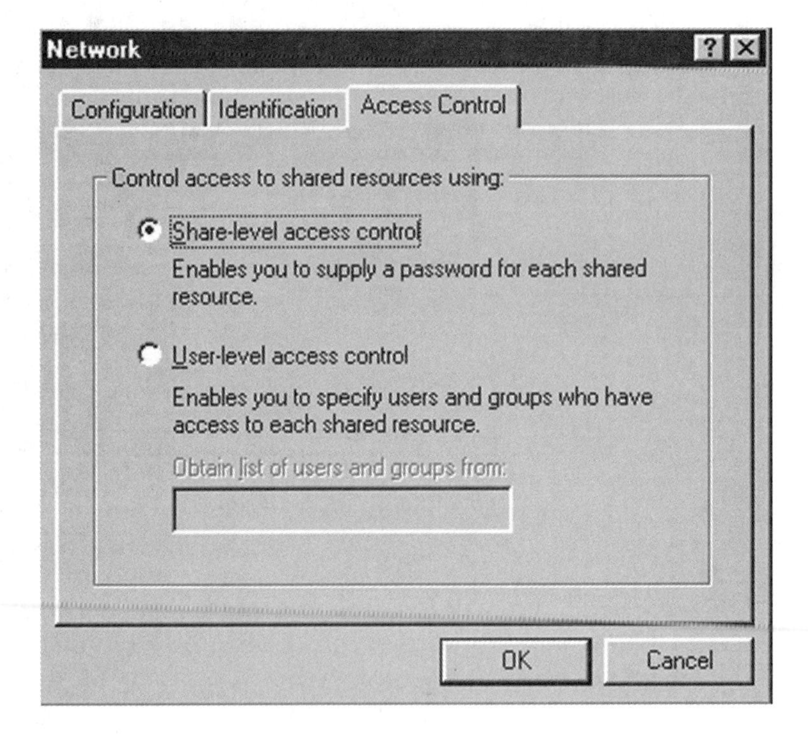

Figure 16-6 Share-Level Access

3. Setting up sharing and security.

   A. Launch Windows Explorer by either clicking on **Start/Programs/Windows Explorer** or by alternate clicking on **Start/Explore.**

   B. In Windows Explorer, make a folder named Test on the C drive and put in four Word documents. See Figure 16-8.

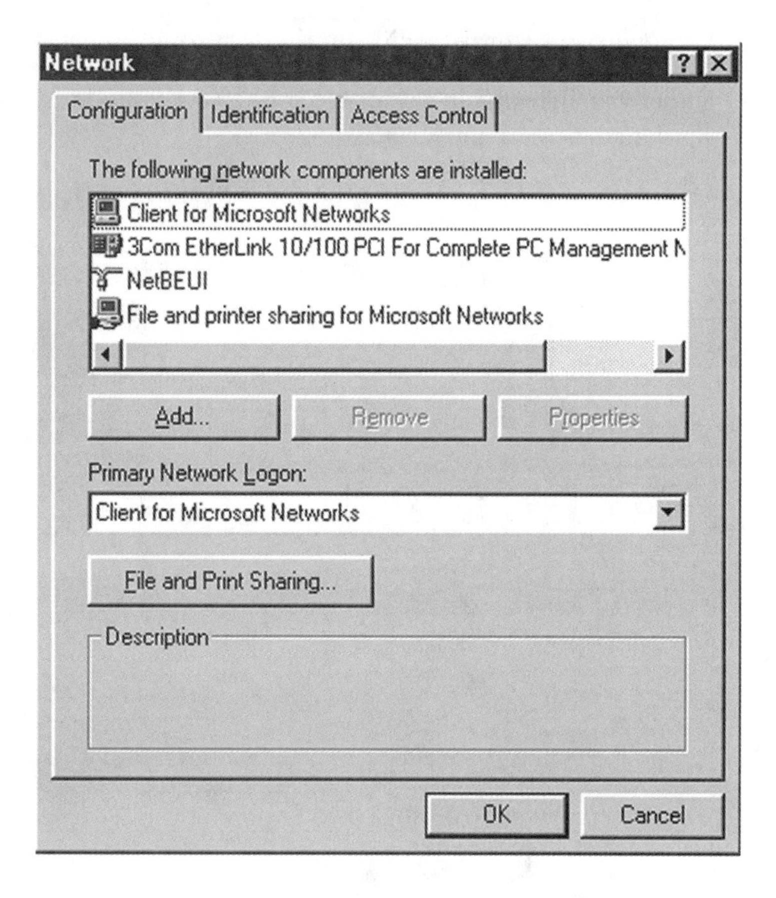

Figure 16-7 Configuration

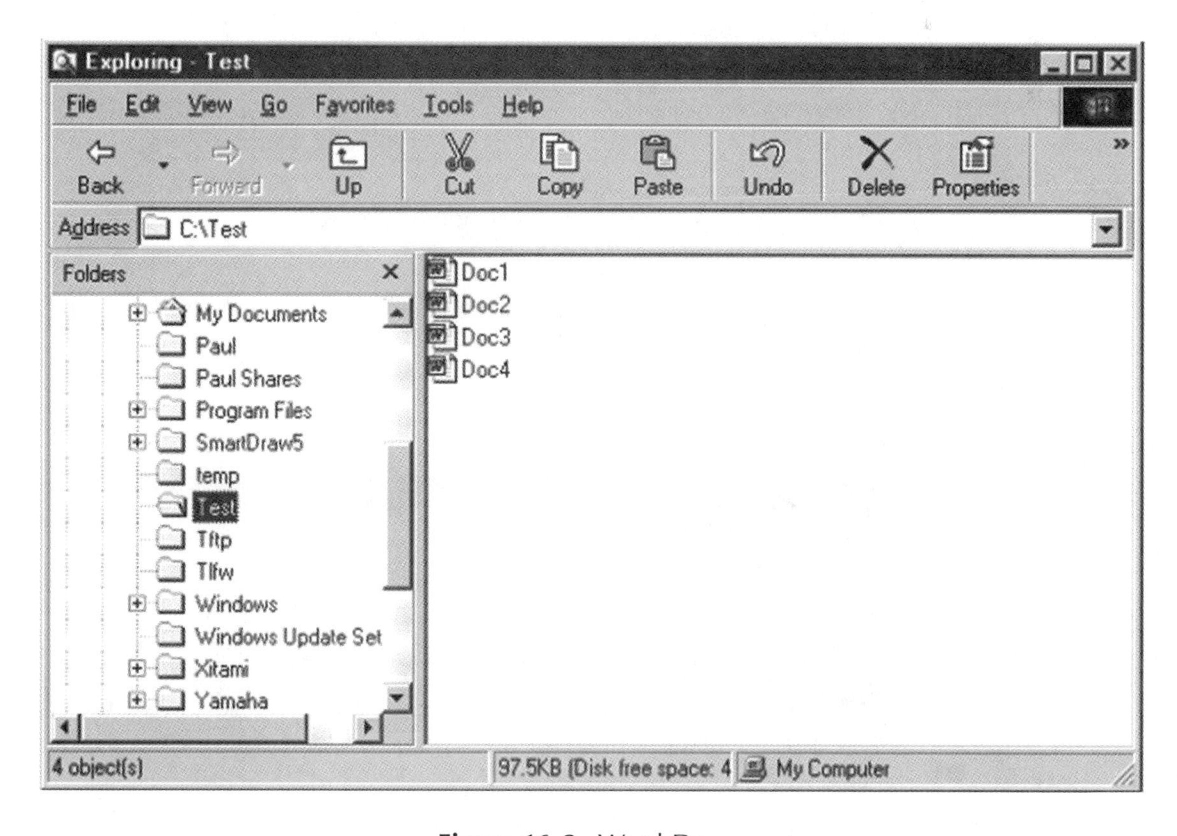

Figure 16-8 Word Documents

C. Highlight and alternate click on the **Test** folder then select **Sharing** and click.

D. Under the Sharing tab select the **Shared As** radio button and enter **TEST** as the share name. See Figure 16-9.

E. Under the Access Type select the radio button for **Depends on Password**.

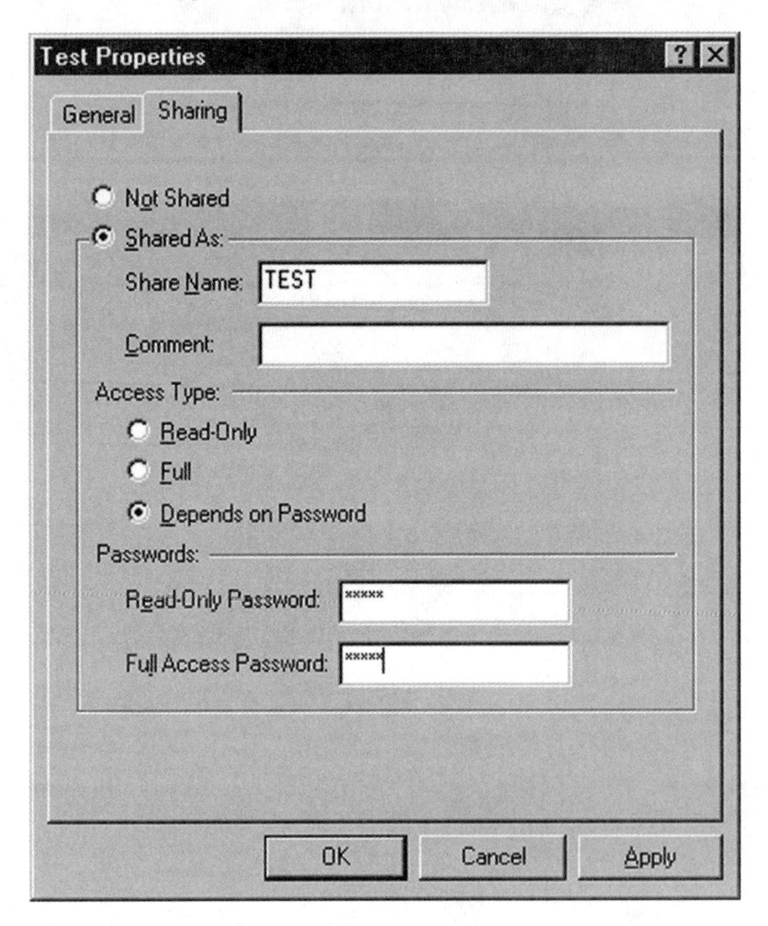

Figure 16-9 Depends on Password

F. Under Passwords enter **test1** for the Read-Only Password and **test2** for the Full Access Password. See Figure 16-9.

G. The program will then ask you to confirm your passwords by reentering them. Click **OK** when done. See Figure 16-10.

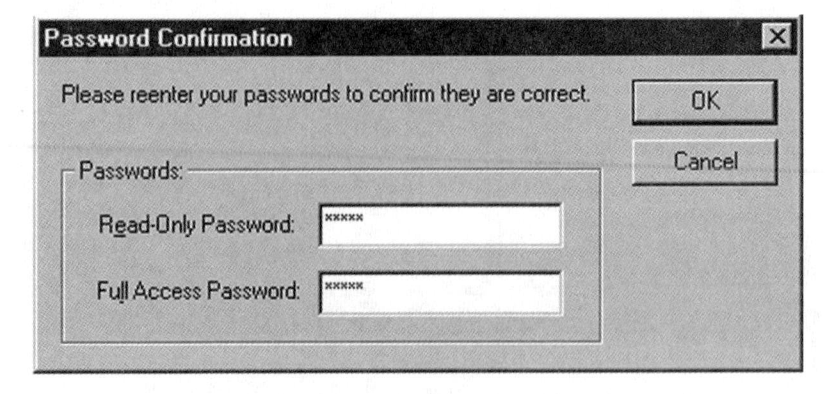

Figure 16-10 Reenter Passwords

4. Testing your peer-to-peer network.

   A. From your desktop open **Network Neighborhood**.

   B. Double-click the other computer on your network and you should see all the shared resources on that computer.

   C. Verify that the TEST folder appears.

   Verified: ___________

   D. Double-click the **TEST** folder and a screen should appear asking for the password. See Figure 16-11.

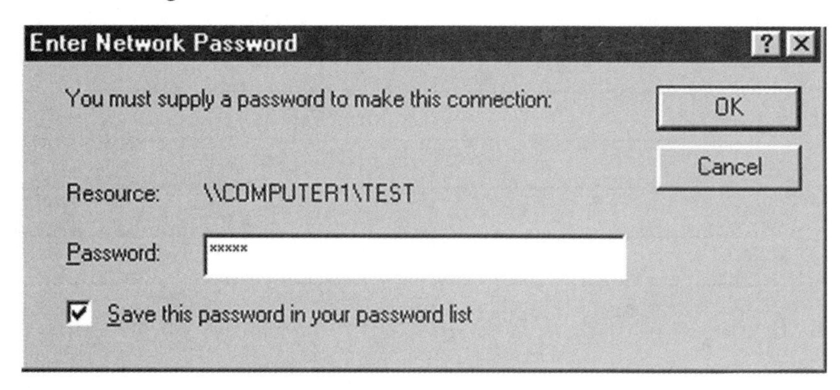

Figure 16-11 Password Screen

   E. Enter **test1** for the password for Read-Only Access.

   F. Open a document, modify it, and then try to save it in the TEST folder.

   G. What was the result?

   ____

   ____

   ____

   H. Shut down your computer and follow steps A through F using **test2** as your password.

   I. What was the result?

   ____

   ____

   ____

5. Try sharing some other resources such as a CD-ROM reader or a printer. Remember that in order to share a printer, print sharing must be enabled in Network Properties.

Questions

1. Are passwords case sensitive when sharing folders?

2. What is the difference between Read-Only and Full Access?

3. If you do not log in, can you access shared files on another computer?

4. NetBEUI makes use of what layers in the OSI?

Experiment 17

Protocol Analysis

Name ____________________ Class __________ Date __________

Objectives Upon completion of this experiment, you should be able to:

- Install a protocol analyzer on your computer.
- Use a protocol analyzer to view packets on a computer network.
- Use the ping command.
- Understand the ICMP packet.

Text Reference Snyder, *Introduction to Telecommunications Networks*
Chapter 10, Section 1

Materials and Equipment
Ethereal Protocol Analyzer
Computer with a Windows OS
NIC
Ethernet Network
Internet Access

Introduction

Ethereal is a protocol analyzer originally authored by Gerald Combs that is available for free at http://www.ethereal.com. Ethereal can be run on either a Windows or Unix-type platform. With Ethereal you can view real-time packet data from a network. In this experiment you will generate network traffic by using the ping command and will view the results on the protocol analyzer. The ping utility is an administration tool that is used to see if a computer is operating and connected to a network. Ping uses the Internet Control Message Protocol (ICMP) Echo function, which returns a message to the sender about its connectivity.

Procedures

1. Download and install the Ethereal program from the Ethereal Web site.
2. The file name is *ethereal-setup-0.9.7.exe* and the file size is 8.384 MB. Ethereal needs another file to make it work, WinPcap, which may or may not be on your computer. If you need it you will have to download it from http://winpcap.polito.it/. WinPcap is a packet-filter device driver for Win32 platforms.
3. Gather information about your computer.

   A. You will need to identify who you are so that when you look at network data you can tell what traffic is being generated by your computer.

B. Network card name ___________ (from Network Properties)

C. Computer name ___________ (from the Identification tab in Network Properties)

D. IP address ___________ (obtained by running c:\>winipcfg/all)

E. Physical address ___________ (obtained by running c:\>winipcfg/all)

F. IP address from another computer near you ___________ (obtained by asking your neighbor)

4. Ethereal setup.

   A. Before using Ethereal or any "sniffing" program, always ask permission to do so from the Network Administrator.

   B. Launch the Ethereal program. See Figure 17-1.

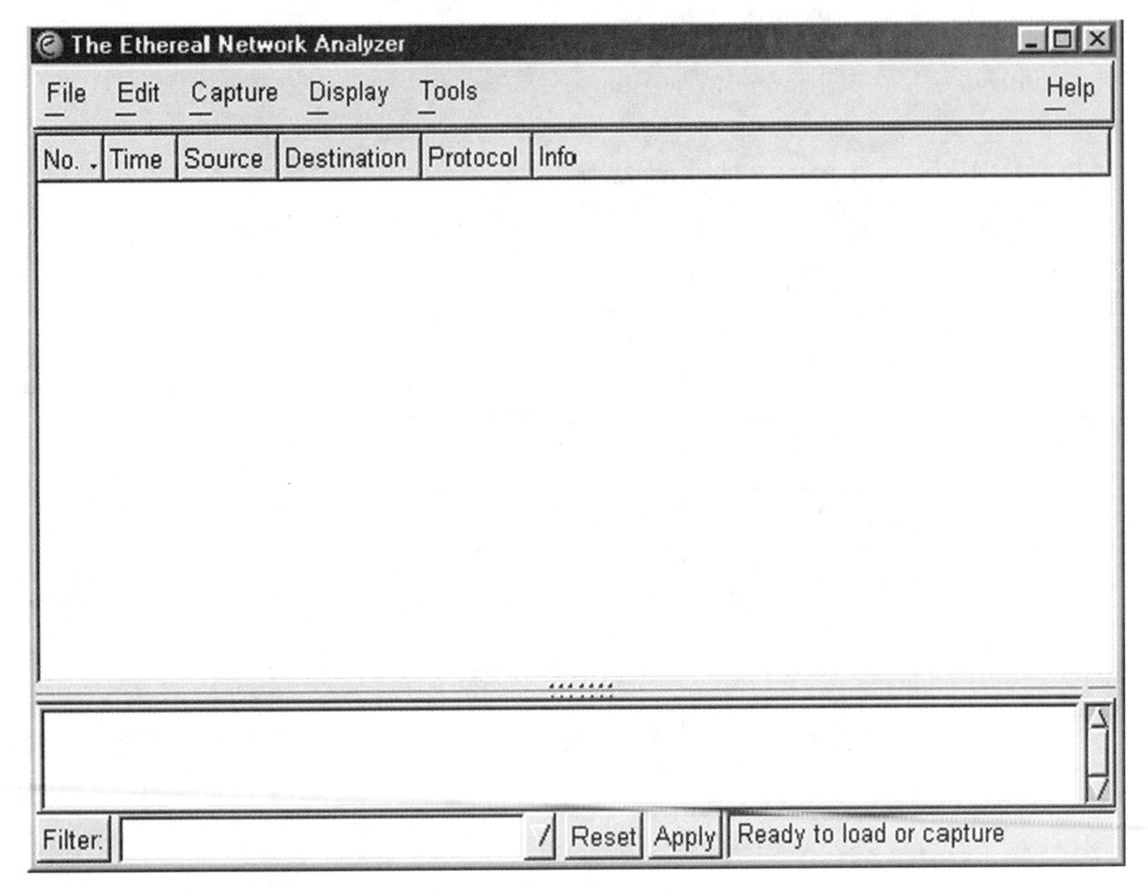

Figure 17-1 Ethereal Main Screen

   C. Go to **Display** and then to **Options**. See Figure 17-2.

      (1) Select **Automatic scrolling in live capture.**

      (2) Select **Enable MAC name resolution.**

      (3) Select **Enable network name resolution.**

      (4) Select **Enable transport name resolution.**

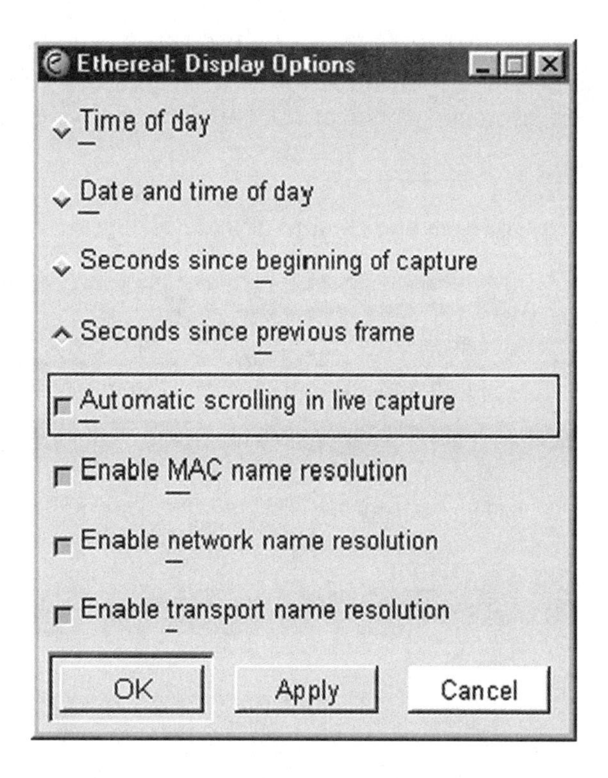

Figure 17-2 Ethereal Display Options

D. Go to **Edit** and then to **Protocols.** See Figure 17-3.

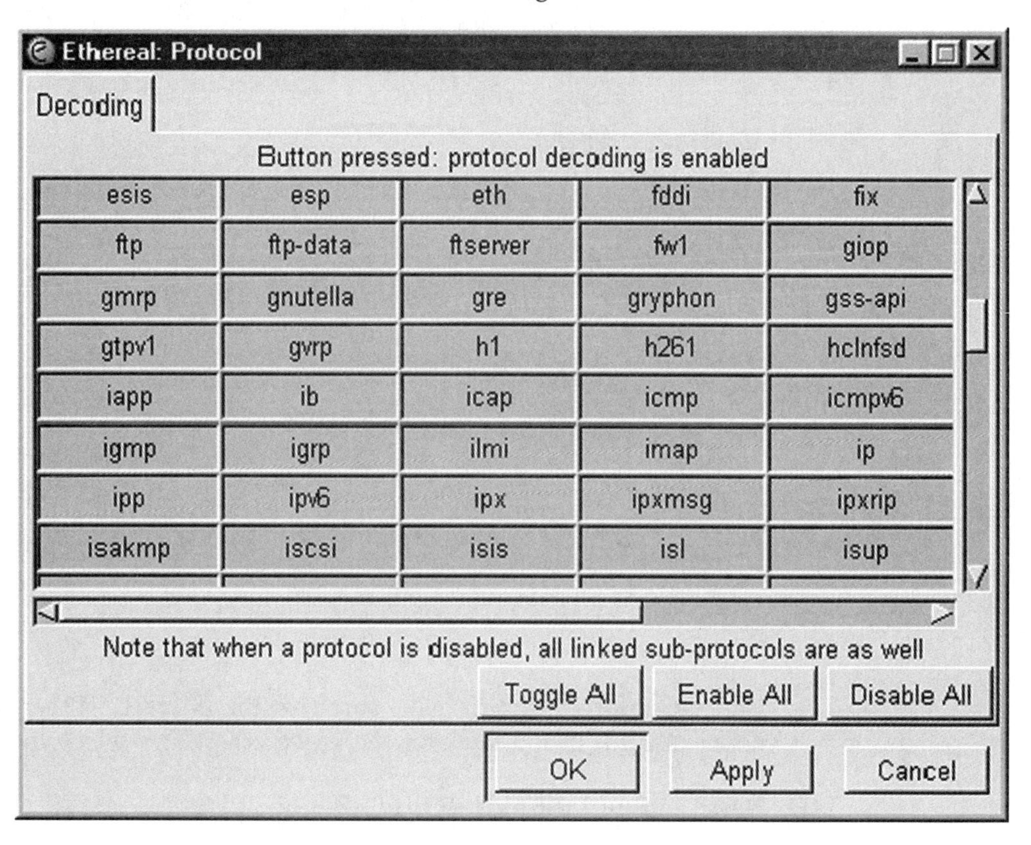

Figure 17-3 Ethereal Protocol

E. View the protocol list and find ICMP. Put the mouse arrow over it and it will display the full meaning of the acronym. Try a few others and then count how many protocols Ethereal can capture.

Count = ___________

F. Go to **Capture** and then to **Start.** See Figure 17-4.

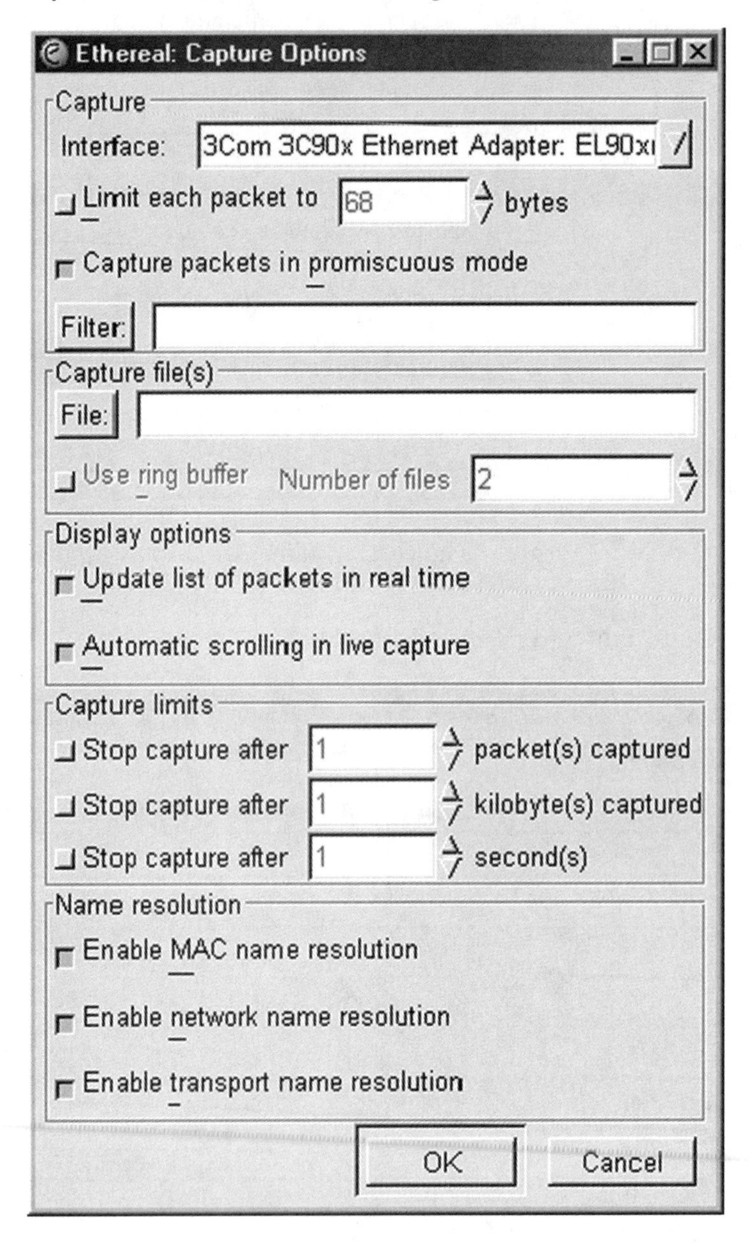

Figure 17-4 Ethereal Capture Options

1. Under **Capture** check to make sure that the interface is that of your network card. If not, click on the down arrow and select your card.
2. Select **Capture packets in promiscuous mode.**
3. Under **Display options** select **Update list of packets in real time.**
4. Click **OK** to start Capture.

5. Generating ICMP packets.

   A. Open a DOS window and ping your neighbor's address by typing **c:\>ping** (address from step 3F).

   B. You should get four replies with some timing information.

      Verified: ___________

   C. Ping another address that does not exist.

   D. You should get four replies that say "Request timed out."

      Verified: ___________

   E. To stop traffic capture click **Stop** in **Ethernet Capture**.

6. Viewing protocol information.

   A. As you can see, you have picked up all the traffic on the wire.

   B. Click and highlight a **captured packet** in the **Ethereal** window and view the headers of the captured traffic. See Figure 17-5.

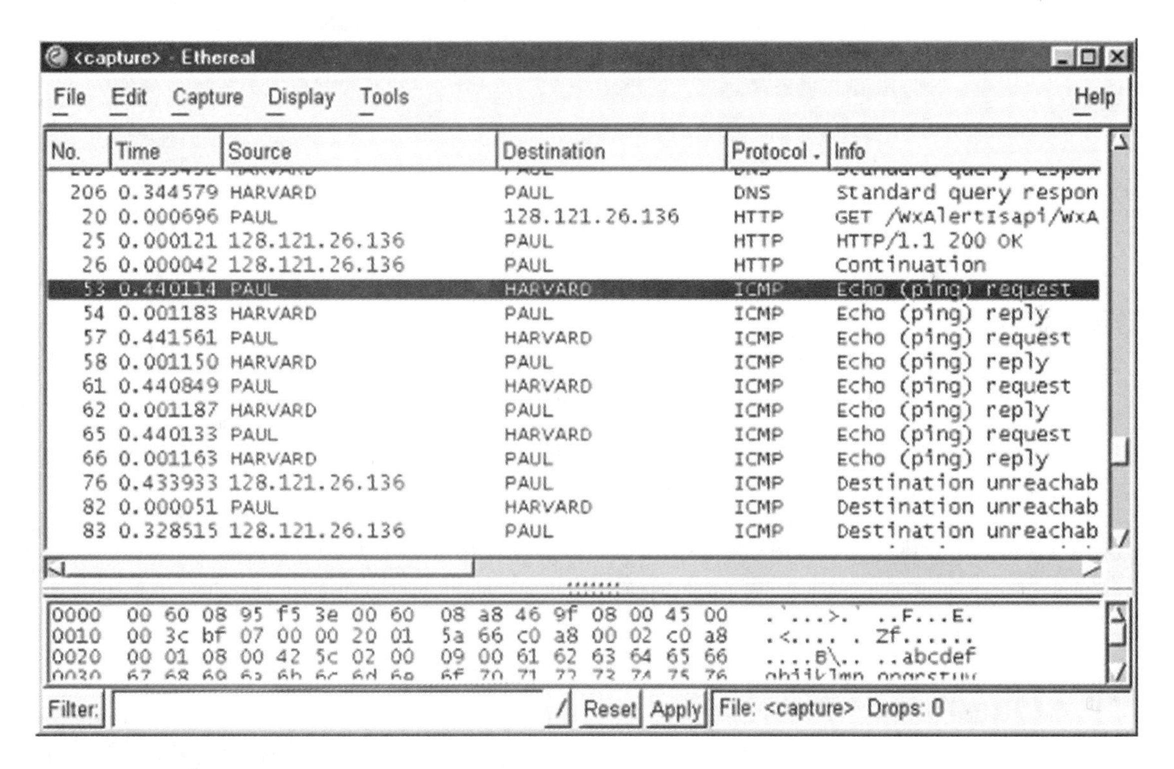

Figure 17-5 Ethereal Capture

   C. You can sort out the protocols by clicking on **protocol**, which will put them in alphabetical order.

   D. Select one of the ICMP (ping) reply packets, expand the center window to analyze the packet in more depth. See Figure 17-6.

   E. Notice that the IP address of both the sending and receiving computer is listed along with the MAC address and the computer name.

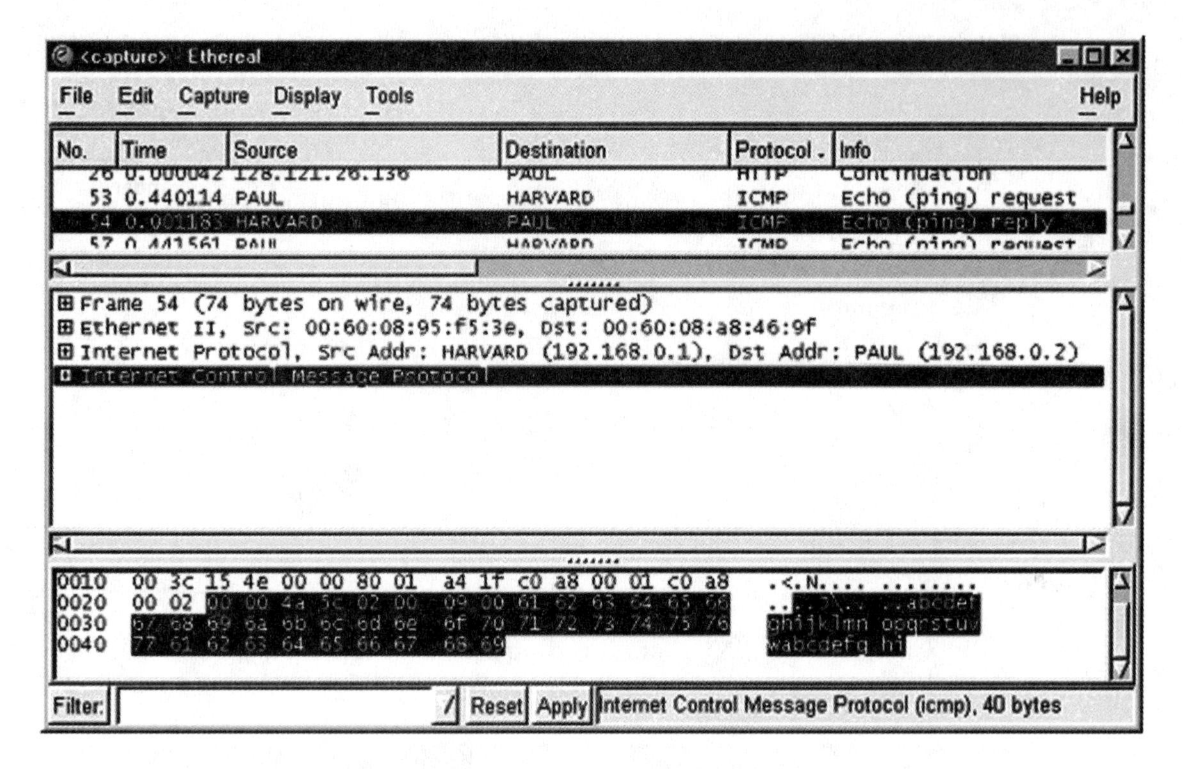

Figure 17-6 Ethereal ICMP

F. Clicking on the plus sign expands the subcategory within the packet to give you more detailed information.

G. Some other protocols to look at are HTTP for those people accessing the Internet. See Figure 17-7.

H. Notice that it gives you information about the Web site that was requested and from which computer the request originated.

I. Have some classmates access some Web sites and see if you can figure out where they go and who they are.

Questions

1. How do you think this tool could help you in troubleshooting a network?

____

____

____

____

____

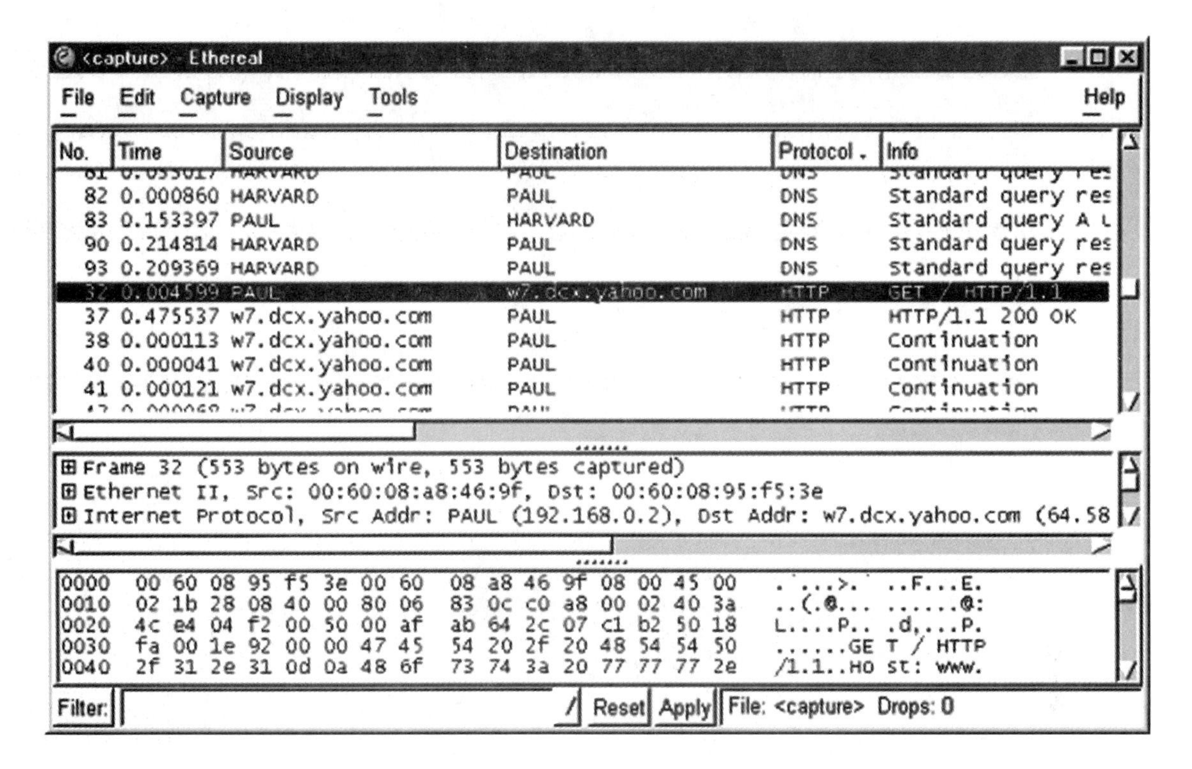

Figure 17-7 Ethereal HTTP

2. What port does HTTP use?

____

____

____

____

____

3. How big was the average packet that was sent out or received?

____

____

____

____

____

4. What was the average time between packets on the wire?

____

____

____

____

____

Experiment 18

UTP Termination: The 568A Standard

Name ________________________ Class ____________ Date ______________

Objectives Upon completion of this experiment, you should be able to:

- Terminate a UTP cable utilizing the EIA/TIA 568A wiring pattern.
- Check cable integrity with a cable tester.
- Connect to a network using your cable.

Text Reference Snyder, *Introduction to Telecommunications Networks*
Chapter 10, Section 3

Materials and Equipment
Category-5 Wire, 4 Pair
RJ-45 Plug (2)
Wire Stripper
Wire Cutter
RJ-45 Crimping Tool
Fluke 620 LAN Cable Meter
Computer with Network Card

Introduction

Cabling is a very important component of a network. Technicians are often called upon to install, troubleshoot, and repair wiring in computer networks. For optimal network performance, the correct wire and termination must be installed. EIA/TIA 568A is the current LAN wiring and termination standard.

Procedures

1. Wire termination straight through cable.

   A. View the RJ-45 plug and notice how, when crimped, the plug will press the connectors through the insulation and make contact with the copper wire. Also notice that the strain relief presses into the outer jacket of the wire to hold it in place. See Figure 18-1.

   B. Cut a piece of CAT-5 cable of a length that will allow you connect a computer to a network (one meter minimum).

   C. Strip 3 in. off the jacket.

   D. Untwist the wires.

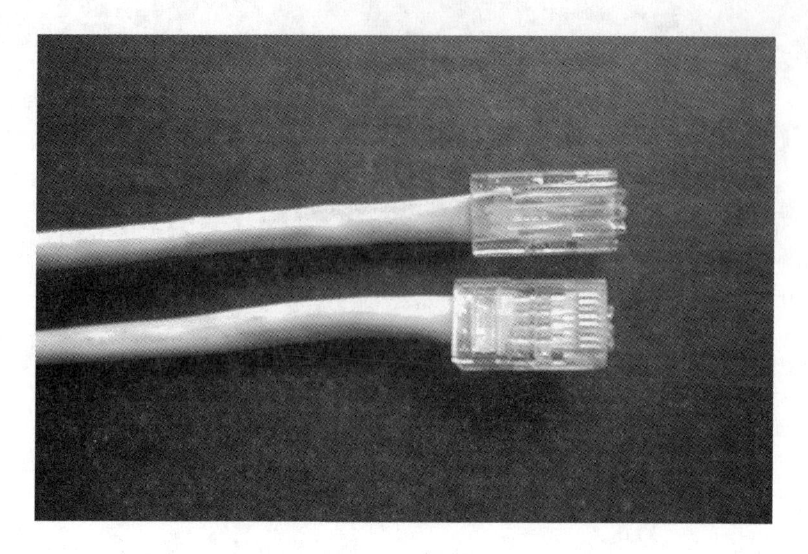

Figure 18-1 Picture of RJ-45

E. Put the wires in order of color scheme. See Table 18-1 and Figure 18-2 for assistance.

F. Flatten and straighten out the wires.

Pin #	Pair	Color
1	3	White/Green
2	3	Green
3	2	White/Orange
4	1	Blue
5	1	White/Blue
6	2	Orange
7	4	White/Brown
8	4	Brown

Table 18-1

G. Cut to 1/2 inch from the edge of the jacket as shown in Figure 18-3.

H. Do *not* strip the wires.

I. Push the wires into the plug in the correct order. Double-check your sequence.

Verified: ___________

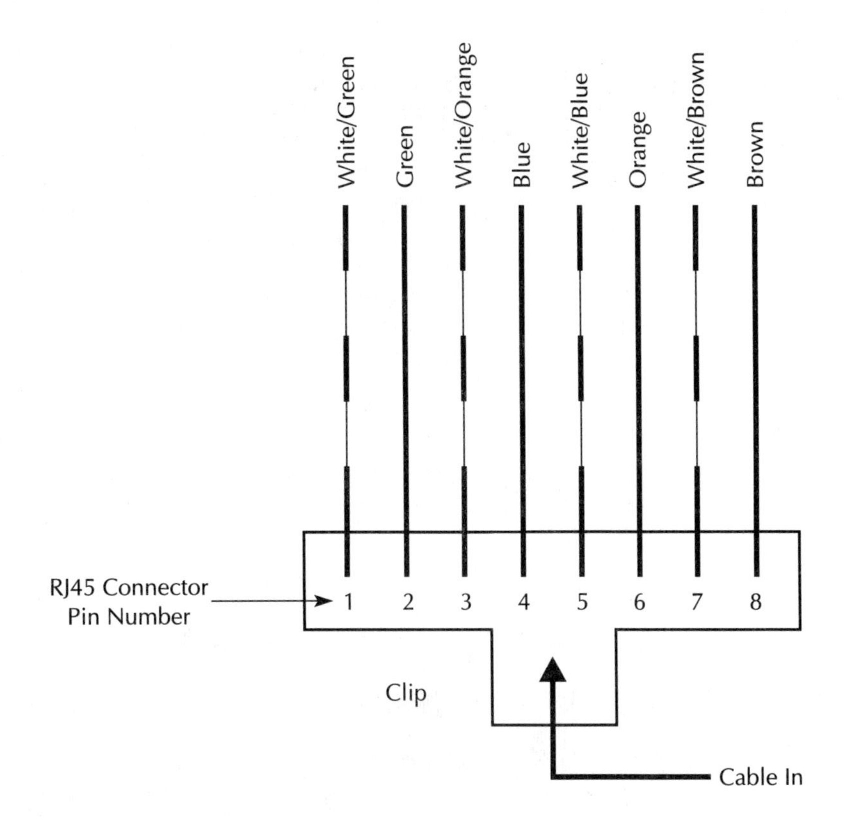

Figure 18-2 Color Scheme 568A

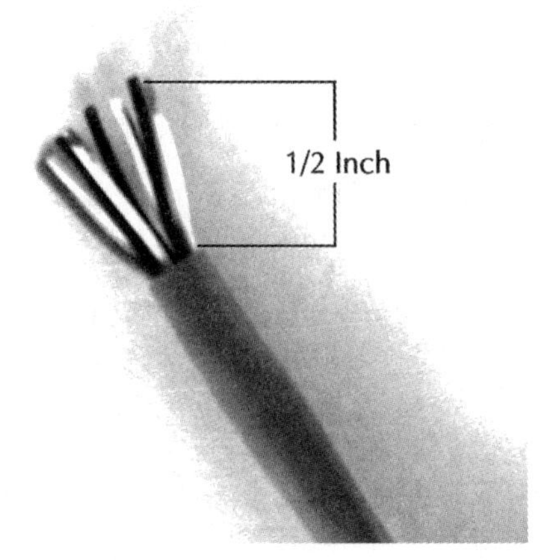

Figure 18-3 Cut Wire to 1/2 Inch

J. Before crimping, look at the end of the plug and verify that the wire goes to the end of the plug. You should see the copper, as depicted in Figure 18-4.

Verified: ____________

K. Crimp the connector using the RJ-45 crimping tool shown in Figure 18-5.

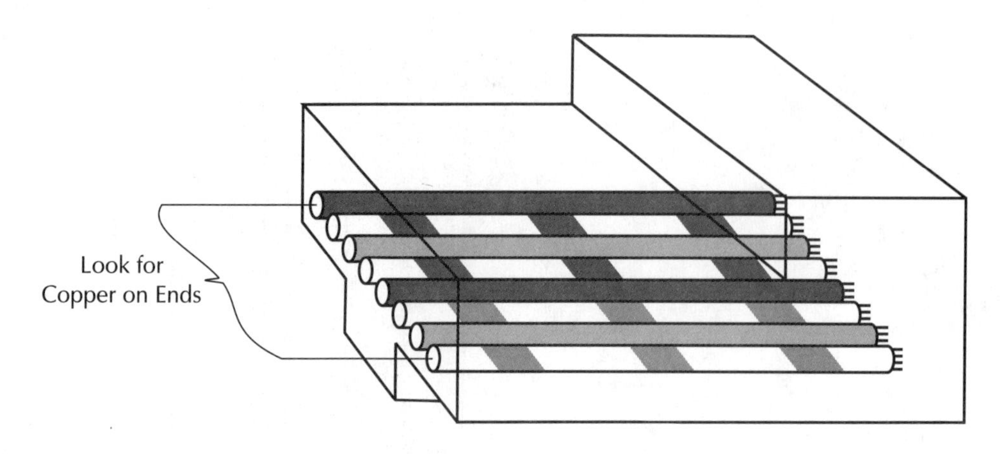

Figure 18-4 Picture of RJ-45 with Wires on End

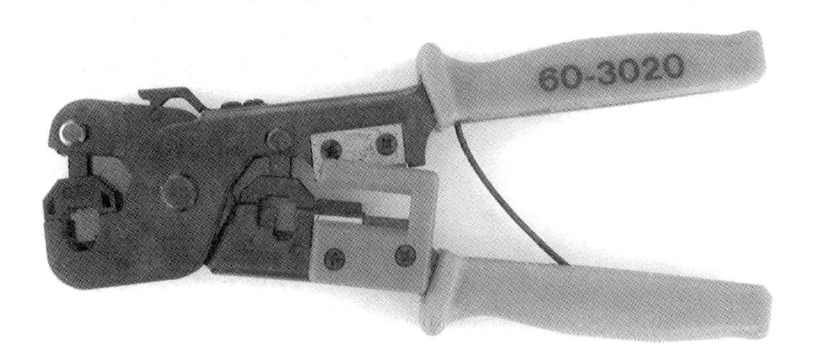

Figure 18-5 RJ-45 Crimper

2. Check cable integrity with a cable tester.

   A. Read the instruction manual for proper operation of the cable tester. See Figure 18-6 for a picture.

   B. The cable tester can test for the following errors:

      (1) Miswire

      (2) Open wire

      (3) Different pail length

      (4) Split pair

   C. Use a cable tester to test the cable.

   D. Is the wire properly terminated? ___________

   E. If yes, what is the length? ___________

   F. If no, then find the fault.

   G. What was the fault?

   ____

   ____

   ____

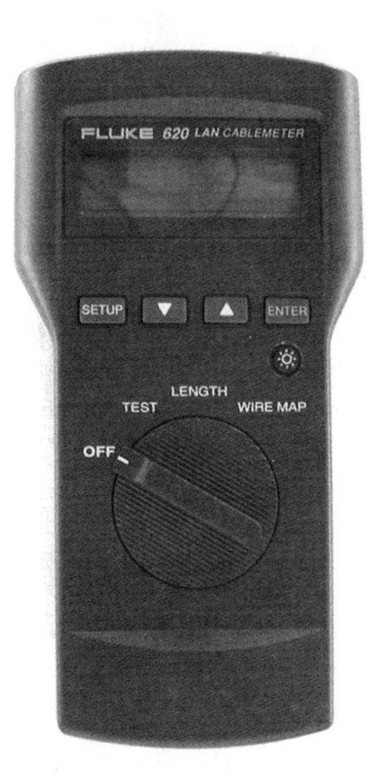

Figure 18-6 Fluke Cable Meter

H. How was the fault determined?

____

____

____

I. Re-terminate and test. (You must have a cable in proper operating condition to go to the next step.)

J. Replace your current network cable with the one you just made and test for network connectivity.

Questions

1. What does RJ stand for in RJ-45?

____

____

____

____

____

2. Who designates the 568A standard?

3. What can you do with this cable?

4. Is it necessary to strip the wires before termination?

5. What is the gauge of the wire used?

6. What is the maximum length of a CAT-5 cable used for Ethernet operation?

7. Which pins are utilized on the RJ-45 connector for Ethernet operation?

Experiment 19

UTP Crossover Cable

Name ______________________________ Class __________ Date ______________

Objectives Upon completion of this experiment, you should be able to:

- Build a crossover cable utilizing an EIA/TIA 568A wiring pattern on one end and an EIA/TIA 568B wiring pattern on the other end.
- Check cable integrity with a cable tester.
- Connect two computers together and check connectivity.

Text Reference Snyder, *Introduction to Telecommunications Networks*
Chapter 10, Section 5

Materials and Equipment
Category-5 Wire, 4 Pair
RJ-45 Plug (2)
Wire Stripper
Wire Cutters
RJ-45 Crimping Tool
Fluke 620 LAN Cable Meter
Computer with Network Card

Introduction

Crossover cables are used to directly connect two similar devices hub to hub, switch to switch, and NIC to NIC. We will be connecting two personal computers together via the NIC cards without the use of a hub or a switch. A crossover cable can be a valuable resource when the only requirement you have is to connect two personal computers together such as a standalone PC and a laptop.

Procedures

1. 568A termination.

   A. View the RJ-45 plug and notice how, when crimped, the plug will press the connectors through the insulation and make contact with the copper wire. Also notice that the strain relief presses into the outer jacket of the wire to hold it in place. See Figure 19-1.

   B. Cut a piece of CAT-5 cable of a length that will allow you connect between the two computers you will be networking (one meter minimum).

   C. Strip 3 in. off the jacket.

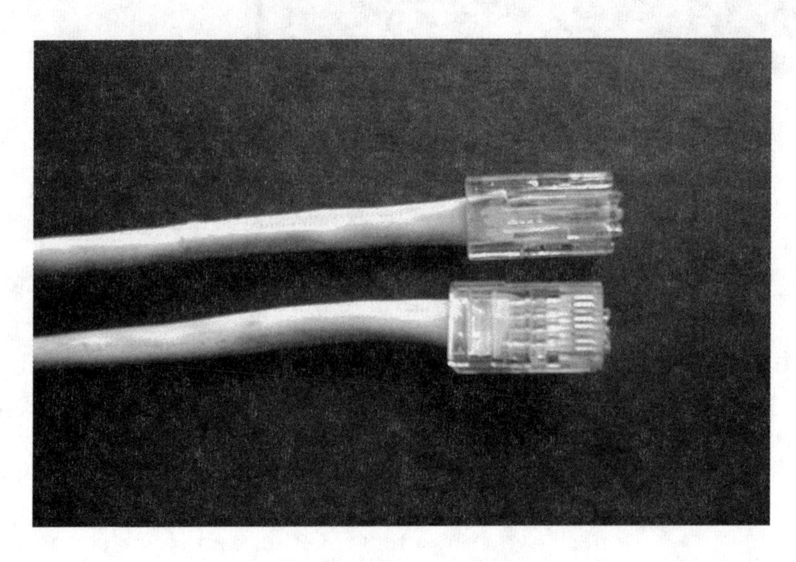

Figure 19-1 Picture of RJ-45

D. Untwist the wires.

E. Put the wires in order of color scheme. See Table 19-1 and Figure 19-2 for assistance.

Pin #	Pair	Color
1	3	White/Green
2	3	Green
3	2	White/Orange
4	1	Blue
5	1	White/Blue
6	2	Orange
7	4	White/Brown
8	4	Brown

Table 19-1 EIA/TIA 568A

F. Flatten and straighten out the wires.

G. Cut to 1/2 inch from the edge of the jacket as shown in Figure 19-3.

H. Do *not* strip the wires.

I. Push the wires into the plug in the correct order. Double-check your sequence.

Verified: ____________

J. Before crimping, look at the end of the plug and verify that the wire goes to the end of the plug. You should see the copper, as depicted in Figure 19-4.

Verified: ____________

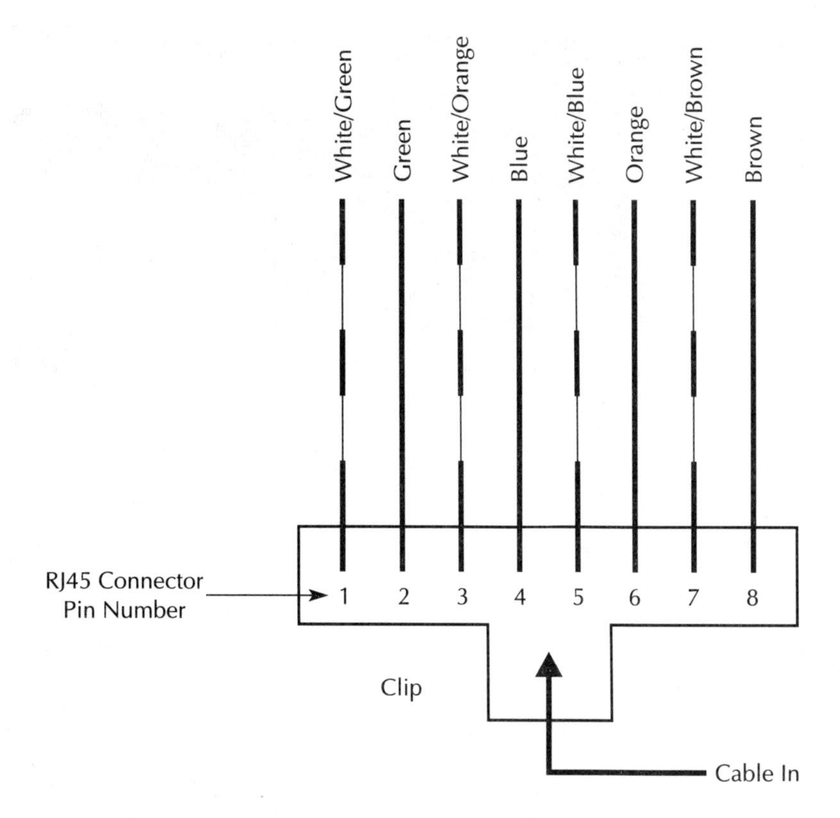

Figure 19-2 Color Scheme 568A

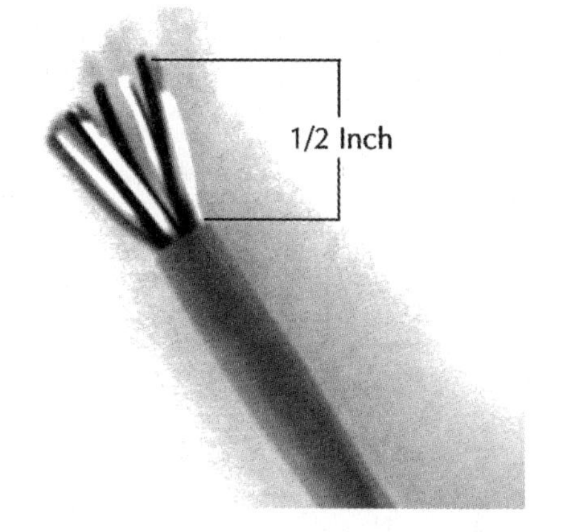

Figure 19-3 Cut Wire to 1/2 Inch

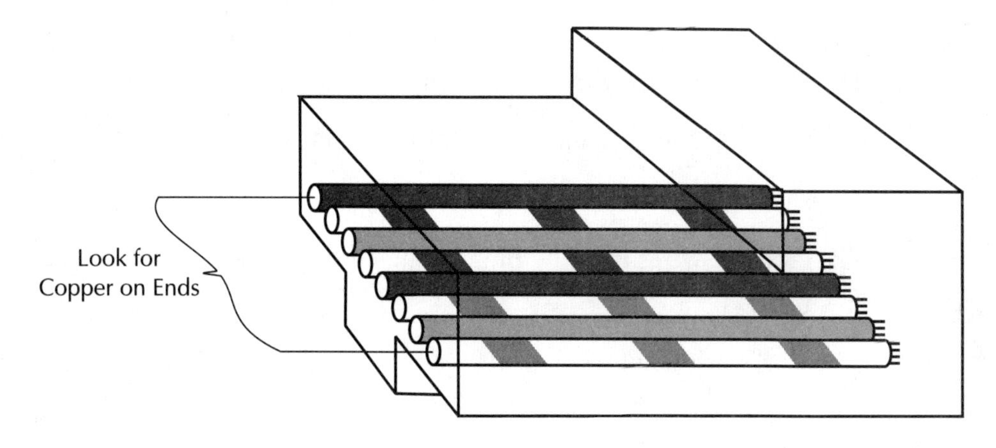

Figure 19-4 Picture of RJ-45 with Wires on End

K. Crimp the connector using the RJ-45 crimping tool shown in Figure 19-5.

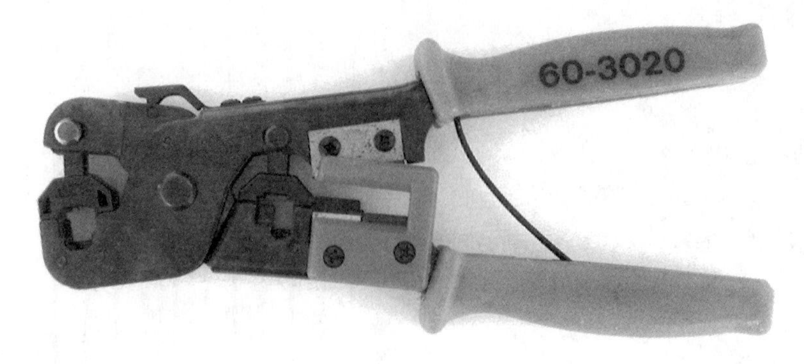

Figure 19-5 RJ-45 Crimper

2. 568B Termination.

   A. Strip 3 in. off the jacket.

   B. Untwist the wires.

   C. Put the wires in order of color scheme. See Table 19-2 and Figure 19-6 for assistance.

Pin #	Pair	Color
1	2	White/Orange
2	2	Orange
3	3	White/Green
4	1	Blue
5	1	White/Blue
6	3	Green
7	4	White/Brown
8	4	Brown

Table 19-2 EIA/TIA 568B

   D. Flatten and straighten out the wires.

   E. Cut to 1/2 inch from the edge of the jacket as shown in Figure 19-3.

   F. Do *not* strip the wires.

   G. Push the wires into the plug in the correct order. Double-check your sequence.

   Verified: ___________

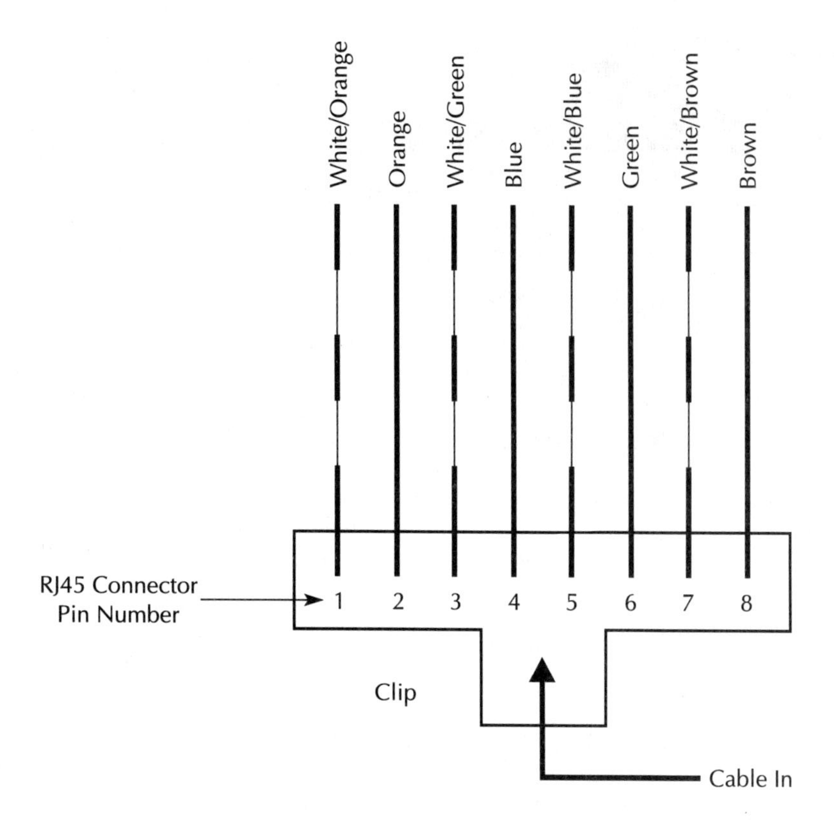

Figure 19-6 Color Scheme 568B

H. Before crimping, look at the end of the plug and verify that the wire goes to the end of the plug. You should see the copper, as depicted in Figure 19-4.

Verified: ___________

I. Crimp the connector using the RJ-45 crimping tool shown in Figure 19-5.

3. Draw a wiring diagram of the cable you made.

4. Read the instruction manual for proper operation of a cable tester.

A. Use a cable tester to test the cable.

B. Does the wire map on the cable tester match the wiring diagram of the cable you made?

Verified: ___________

5. Connect your cable between two computers and test the connectivity.

Questions

1. Why is it important not to untwist more than 1/2 inch of each wire pair?

2. What is another use for a crossover cable?

3. Explain how you tested the cable with the cable tester. Which pairs were crossed?

4. What devices perform the crossover function in an Ethernet network?

Experiment 20

Rollover Cable

Name ______________________ Class ____________ Date ______________

Objectives Upon completion of this experiment, you should be able to:

- Build a rollover cable for connection from a PC to a console port on a switch.
- Check cable integrity with a cable tester.

Text Reference Snyder, *Introduction to Telecommunications Networks*
Chapter 10, Section 5

Materials and Equipment
Category-5 Wire, 4 Pair
RJ-45 Plug (2)
Wire Stripper
Wire Cutters
RJ-45 Crimping Tool
RJ-45 to DB-9 Female Terminal Adapter
Fluke 620 LAN Cable Meter
Computer with Network Card

Introduction

A console cable, also called a rollover cable, is used for connection from a workstation to the console port on a switch or router for the purpose of configuration. The cable uses the serial port on the router or switch, which has an RJ-45 connector, and the workstation, which requires a 9-pin female D connector that plugs into the DB-9 male serial port on the workstation running the HyperTerminal program.

A rollover cable uses 8 pins, but the wiring is different than the straight-through cable or crossover cable. The rollover-cable wiring has pin 1 on one end connect to pin 8 on the other end, pin 2 to pin 7, pin 3 to pin 6, pin 4 to pin 5, pin 5 to pin 4, pin 6 to pin 3, pin 7 to pin 2, and pin 8 to pin 1. Because the pins on one end are reversed, the other end appears to be rotated or rolled over.

The switch comes with an 8-foot rollover cable, but you will make a 12-foot cable so the switch and PC do not have to be so close together.

Procedures

1. RJ-45 termination.

   A. View the RJ-45 plug and notice how, when crimped, the plug will press the connectors through the insulation and make contact with the copper wire. Also notice that the strain relief presses into the outer jacket of the wire to hold it in place. See Figure 20-1.

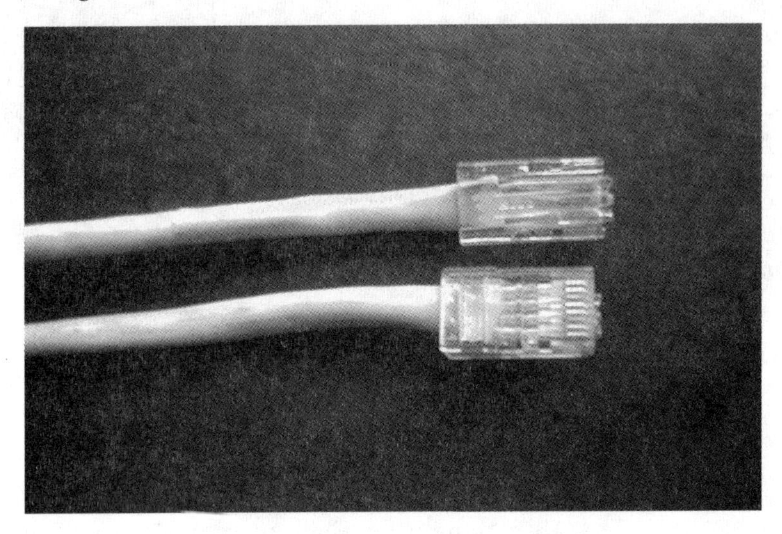

Figure 20-1 Picture of RJ-45

   B. Draw a wire map of the cable you are going to make that shows colors and pin assignments.

   C. Cut a 12-foot piece of CAT-5 cable.

   D. Strip 3 in. off the jacket.

   E. Untwist the wires.

   F. Put the wires in order of color scheme. See Table 20-1 for assistance.

   G. Flatten and straighten out the wires.

   H. Cut to 1/2 inch from the edge of the jacket as shown in Figure 20-2.

   I. Do *not* strip the wires.

   J. Push the wires into the plug in the correct order. Double-check your sequence.

   Verified: ___________

End 1		End 2	
Pin #	Color	Pin #	Color
1	Brown	8	Brown
2	White/Brown	7	White/Brown
3	Green	6	Green
4	White/Blue	5	White/Blue
5	Blue	4	Blue
6	White/Green	3	White/Green
7	Orange	2	Orange
8	White/Orange	1	White/Orange

Table 20-1 Rollover Pin Connections

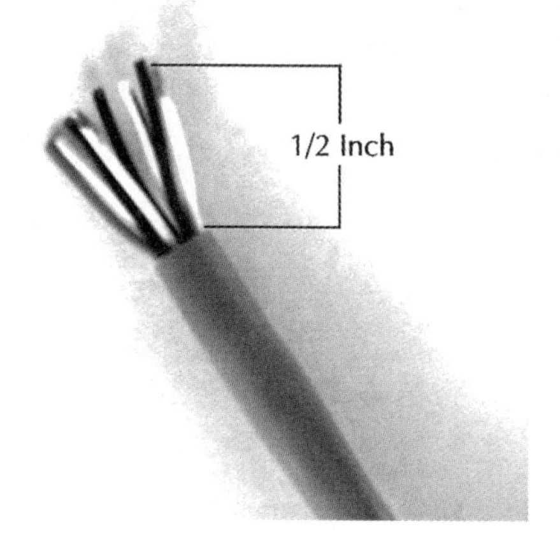

Figure 20-2 Cut Wire to 1/2 Inch

K. Before crimping, look at the end of the plug and verify that the wire goes to the end of the plug. You should see the copper, as depicted in Figure 20-3.

   Verified: ____________

L. Crimp the connector using the RJ-45 crimping tool.

M. Repeat steps D to K for the other end of the cable (End 2).

N. Crimp End 2 using the reverse order from End 1. See Table 20-1.

2. Read the instruction manual for proper operation of a cable tester.

   A. Use a cable tester to test the cable.

   B. Does the wire map match the one you created in step 1B?

      Verified: ____________

3. Label this cable *rollover cable* so you will not get it mixed up with a straight-through cable. Fill in Table 20-2 with the serial port designations, which correspond to the pin number.

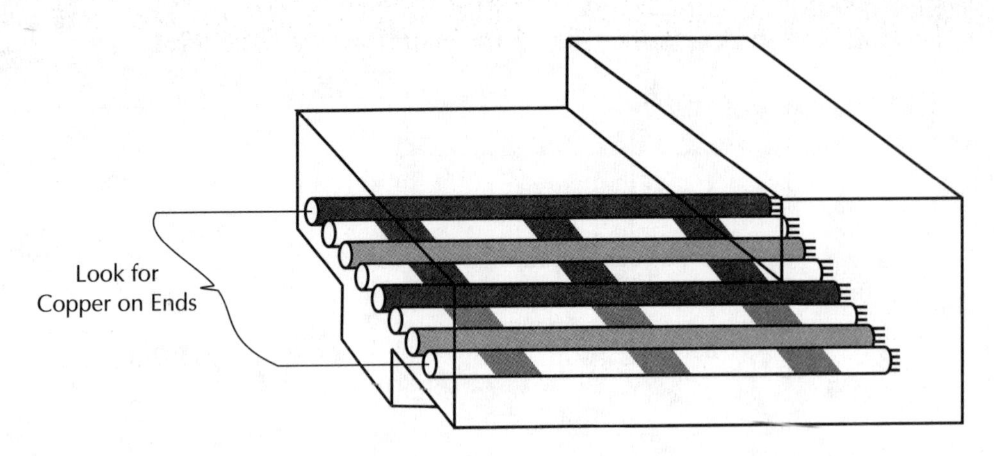

Figure 20-3 Picture of RJ-45 with Wires on End

Pin Number	RS-232 Designation
1	RTS
2	
3	
4	
5	
6	
7	
8	

Table 20-2

Questions

1. How does a rollover cable differ from a straight-through cable in design?

____

____

2. How does a rollover cable differ from a straight-through cable in function?

____

____

3. Which pins are used on a straight-through cable?

____

____

4. Which pins are used on a rollover cable?

____

____

Experiment 21

The Ethernet Switch

Name ______________________ Class __________ Date ____________

Objectives Upon completion of this experiment, you should be able to:

- Identify the types of connections and devices that can attach to a switch.
- Connect to the switch using the HyperTerminal program through the console connection.
- Connect to the switch using the Telnet program.
- Connect to the switch using a Web browser.
- Check and modify the switch configuration.

Text Reference Snyder, *Introduction to Telecommunications Networks*
Chapter 10, Section 5

Materials and Equipment
Cisco Catalyst 1900-Series Switch
Computer with a Free Serial Port
Console (Rollover) Cable
Category-5 Patch Cable from the Computer to the Switch
HyperTerminal Program
Telnet Program
Web Browser

Introduction

A switch is a layer-2 device that acts as the concentration point for the attachment of servers, clients, routers, hubs, and other switches. A switch maintains both a table of hardware addresses and its relationship to the segment it is on, reading the source address of the frame and associating it with the segment. The switch provides a point-to-point connection between the two devices so there are no collisions on that segment.

There are two switching methods that can be used to forward a frame through a switch: store-and-forward and fragment free. Fragment free reads the first 64 bytes of the frame and then forwards the frame. Error correction is low and it exhibits reduced latency. Store-and-forward waits to receive the whole frame before it is forwarded and has better error correction, but it is slower than fragment free.

A switch can be managed either by connecting to the console port with a special cable called a rollover cable or over a LAN using the Telnet program (which will allow you to view and make changes to the configuration). The ability to understand and configure switches is essential for network support.

Procedure

1. Attaching to the switch using the console cable.

   A. View the console cable and notice the differences between it and a network cable that attaches a PC to a switch. Describe the differences.

   ____

   ____

   ____

   ____

   ____

   B. Attach the console cable from the serial port on the PC to the console port on the switch.

   C. Launch the HyperTerminal program and name the connection and click **OK**. See Figure 21-1.

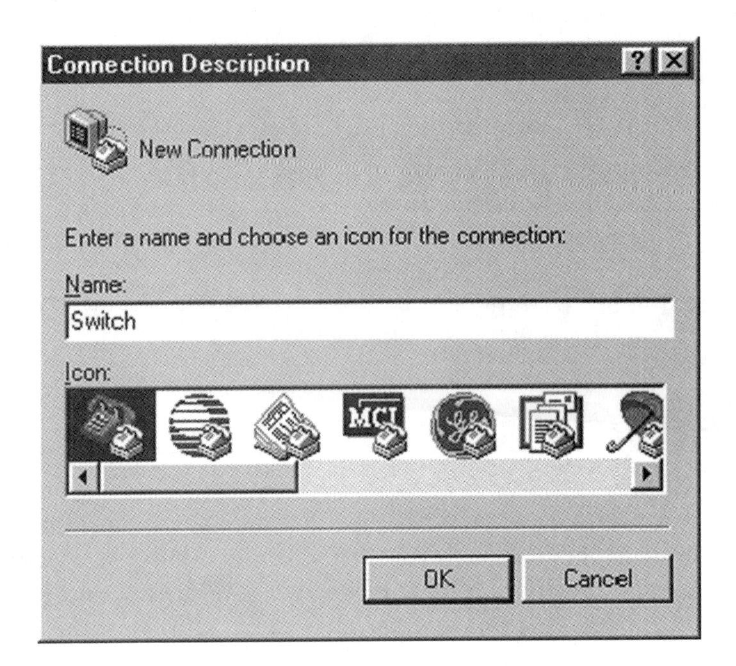

Figure 21-1 HyperTerminal

   D. Connect to the Com port you are using. See Figure 21-2.

   E. Set the properties as detailed in Figure 21-3 and click **OK**.

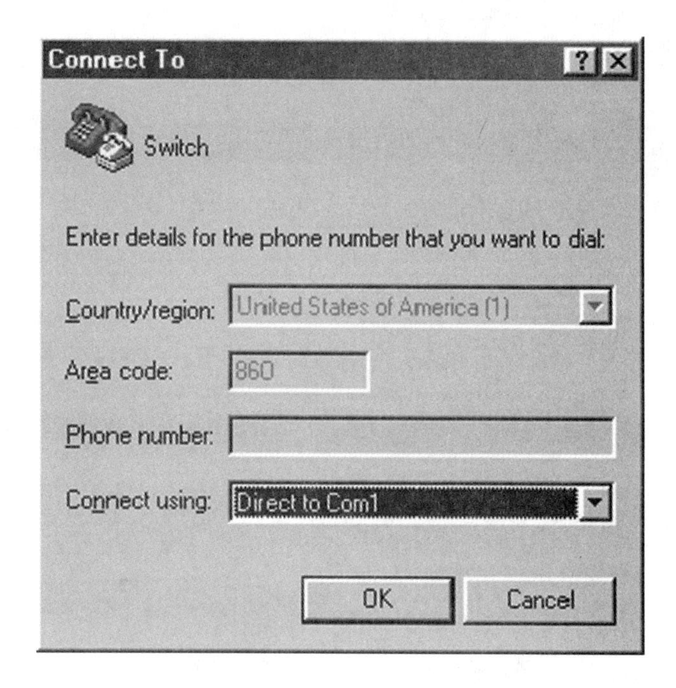

Figure 21-2 Com Port

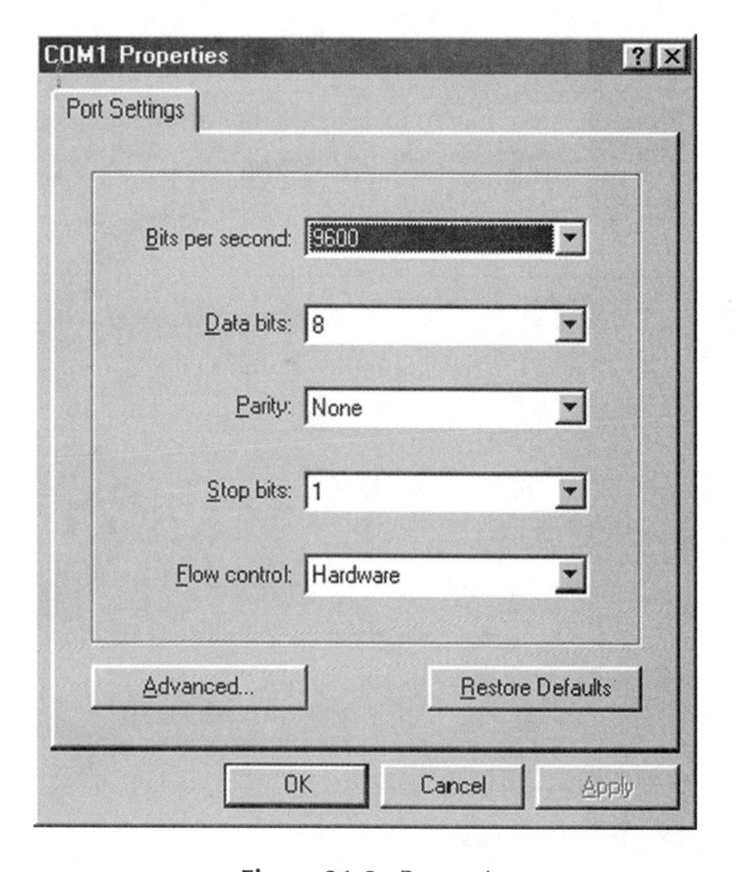

Figure 21-3 Properties

F. Enter the password if one is required (supplied by the instructor). See Figure 21-4.

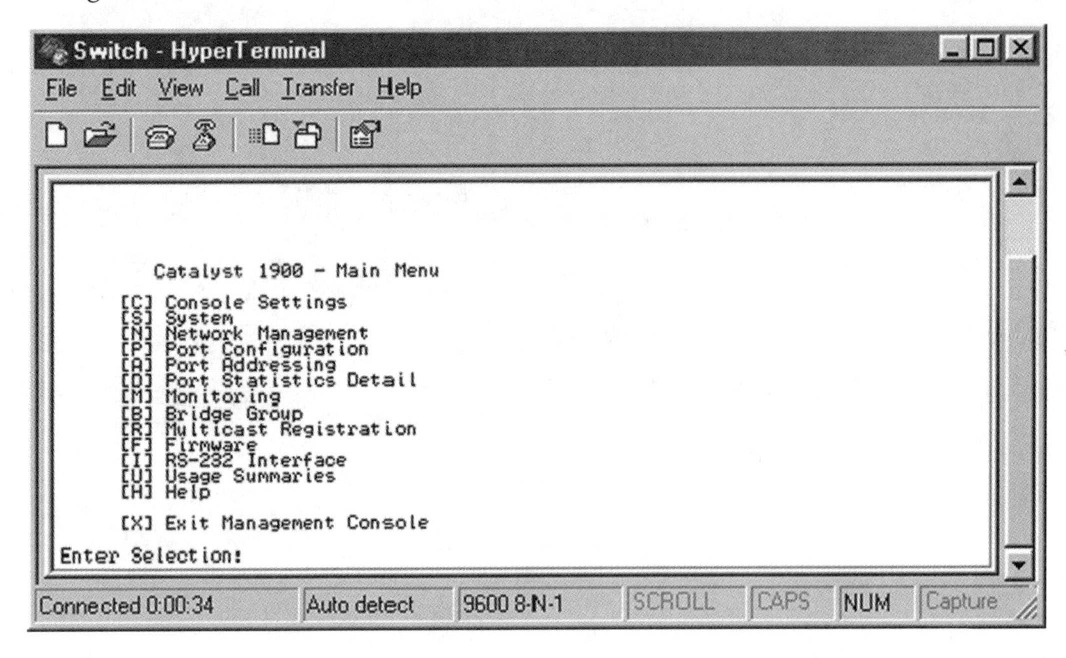

Figure 21-4 Password

G. From the main menu, investigate the different submenus and answer the following, providing the menu where you found the information:

(1) What is the model number of the switch?

____

____

(2) What is the serial number of the switch?

____

____

(3) What is the system name?

____

____

(4) What type of switching mode is used?

____

____

(5) What is the IP address?

____

____

(6) What is the MAC address?

____

____

(7) Is HTTP enabled? If not, enable it.

____

____

(8) What port is it on?

____

____

H. Make sure the IP address and the SM and default gateway match your network's addressing scheme. If they do not, change them to match, as shown in Figure 21-5.

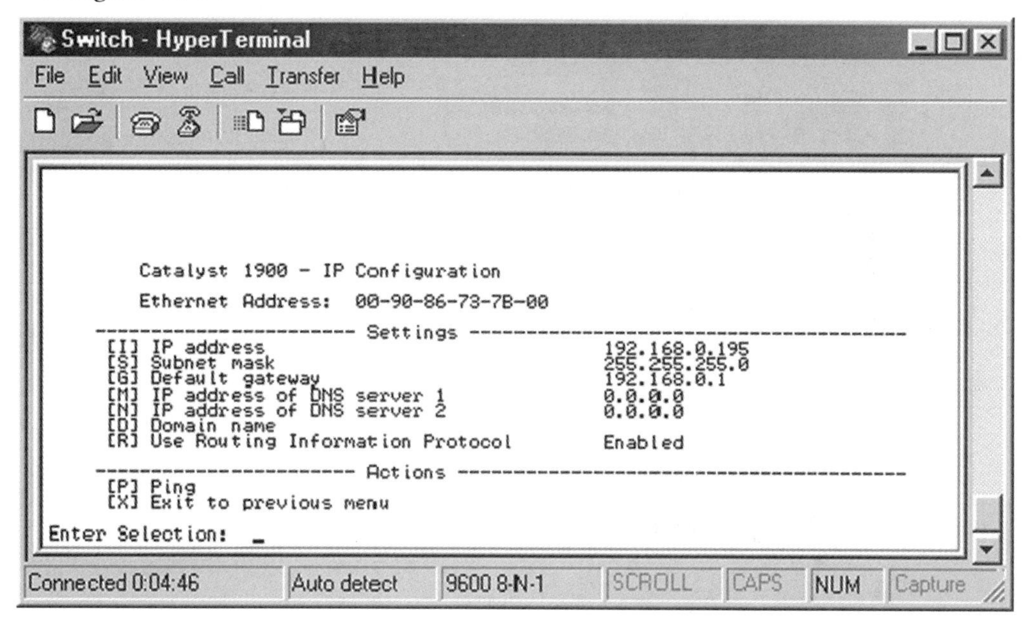

Figure 21-5 IP Address

2. Attach to the switch using the Telnet program.

A. Attach a network cable from the PC to one of the ports on the switch.

B. Start the Telnet program found in the Windows directory.

C. From **Connect**, enter the IP address of the switch and connect as shown in Figure 21-6.

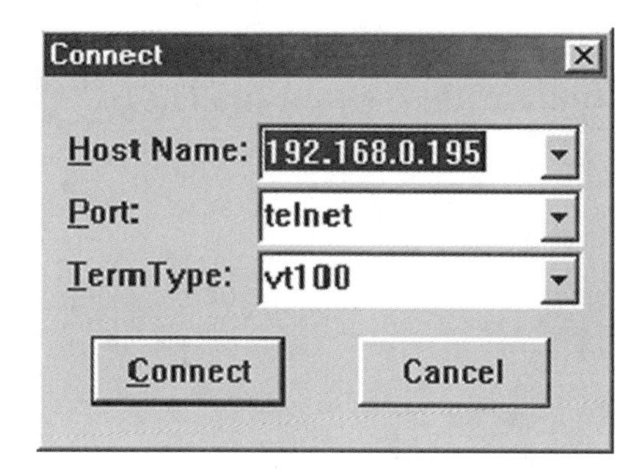

Figure 21-6 Connect

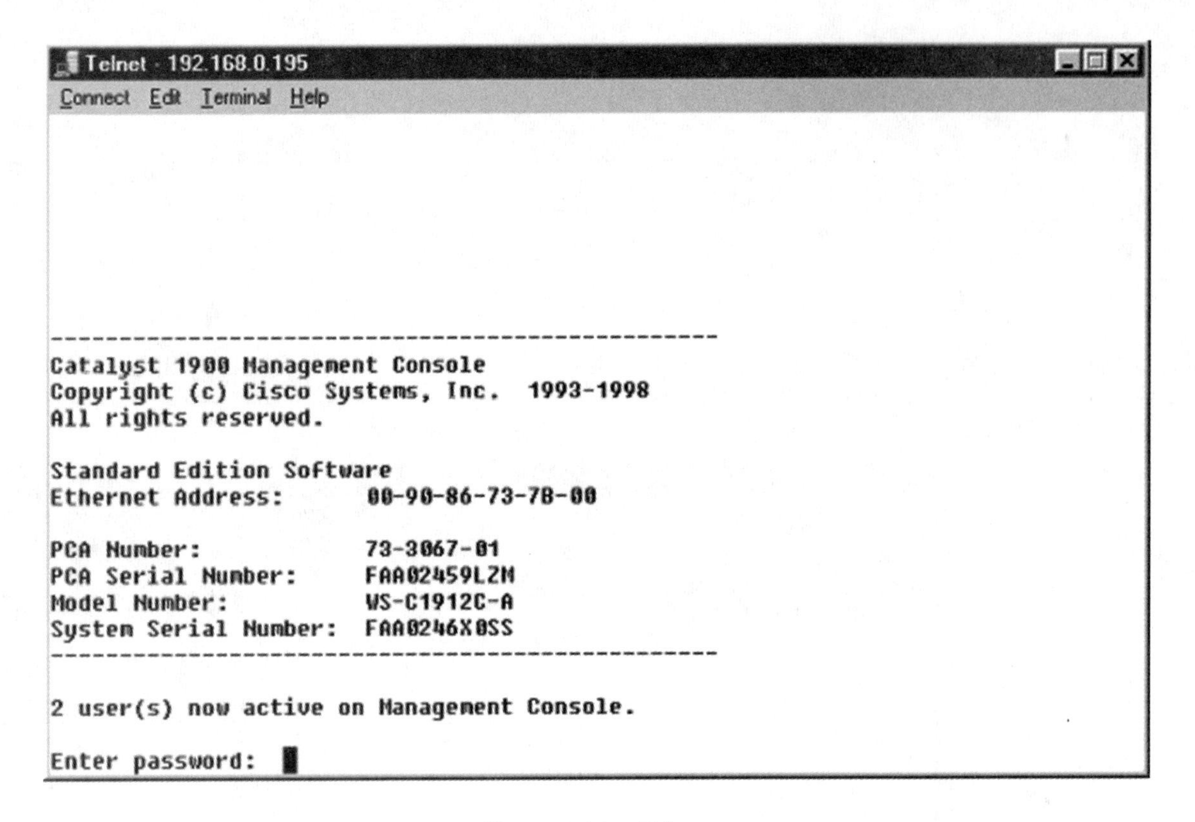

Figure 21-7 Telnet

D. A menu similar to the HyperTerminal menu should be displayed. See Figure 21-7.

E. Enter the password.

F. What are the advantages of using Telnet over HyperTerminal?

____

____

____

____

____

3. Connecting to the switch using a Web browser.

A. Start your Web browser.

B. In the **Address** bar, enter the IP address of the switch.

Questions

1. Is there an ON/OFF switch for power?

2. How do you turn the switch on?

3. How many devices can be attached to the switch?

4. Are there any crossover ports?

5. How can you tell if a port is a crossover port?

6. What does the Mode button do on the front of the switch?

____

____

____

____

____